ICT for Smart Grid - Recent Advances, New Perspectives, and Applications

Edited by Abdelfatteh Haidine

Published in London, United Kingdom

ICT for Smart Grid - Recent Advances, New Perspectives, and Applications
http://dx.doi.org/10.5772/intechopen.1000349
Edited by Abdelfatteh Haidine

Contributors
Abdelfatteh Haidine, Anna Jarosz, Asmaa El Hannani, Bangar Raju Lingampalli, Basma Saad, Bishnu Prasad Muni, Effrain Roney Bernardes, Fatemeh Mechershavi, Gunes Karabulut Kurt, Hassan Ouahmane, Houda Daki, Javad Soleimani, Kiran Kumar Malligunta, Mallu Rama Prasad Reddy, Peter Conrad, Rahul S. Somalwar, Reza Farhangi, Saraswathula Subramanya Sarma, Steffen Späthe, Subba Rao Kotamraju

First published in London, United Kingdom, 2024 by IntechOpen
IntechOpen is the global imprint of INTECHOPEN LIMITED, registered in England and Wales, registration number: 11086078, 167-169 Great Portland Street, London, W1W 5PF, United Kingdom

British Library Cataloguing-in-Publication Data
A catalogue record for this book is available from the British Library

Additional hard and PDF copies can be obtained from orders@intechopen.com

ICT for Smart Grid - Recent Advances, New Perspectives, and Applications
Edited by Abdelfatteh Haidine
p. cm.
Print ISBN 978-0-85466-533-4
Online ISBN 978-0-85466-532-7
eBook (PDF) ISBN 978-0-85466-534-1

Meet the editor

Dr. Ing. Abdelfatteh Haidine is Professor Habilité at the University of Applied Sciences, El Jadida, Morocco. He is a member of the Laboratory of Information Technologies. His research interests include simulation and evaluation of wireless and mobile networks, as well as design (network planning/migration and resources allocation) and evaluation of communication networks for smart environments, especially smart cities, smart grids, smart maritime ports, and smart environment monitoring. He holds a Ph.D from Technische Universität Dresden, Germany, which focused on the planning and optimization of. communication networks by means of combinatorial optimization and multi-objective optimization/multi-criteria decision-making. He has worked on various European research and industry projects. He has also worked for different companies as a consultant for intelligent network communications, with a focus on the planning of communication solutions for power utilities to build smart grid communication platforms. His tasks involved technical specifications, planning and performance evaluation of communication solutions, and techno-economics aspects for smart grid projects.

Contents

Preface

Electrical energy is a fundamental resource that underpins nearly every aspect of modern life, driving progress, innovation, and human well-being. It is the lifeline for very critical activities, such as healthcare facilities, economic activities and industries, communication and information infrastructures, and more. Therefore, it is critical for the country to guarantee a reliable, secure, stable, economic, and sustainable energy infrastructure. To reach this goal, different approaches have been developed under the umbrella concept of the "smart grid".

A smart grid is an advanced electrical grid system that utilizes modern communication, sensing, and control technologies to improve the efficiency, reliability, and sustainability of electricity generation, transmission, distribution, and consumption. Unlike traditional electrical grids, which primarily deliver electricity in a one-way flow from power plants to consumers, smart grids incorporate bidirectional communication and real-time data exchange between various components within the grid. Different features of a smart grid, include the integration of renewable energy, enhanced grid monitoring and control, demand response and load management, facilitation of distributed energy resources (DERs), automated metering infrastructure (AMI), and so on.

This book treats the subject of smart grid and the aspect of using advanced Information and Communication Technologies (ICT). It includes seven chapters that are organized into two sections. The first section, "Smart Grid in Special Environments", includes the first four chapters, and the second section, "Analysis and Modeling in Smart Grids", includes the final three chapters.

Chapter 1, "Assimilation of Sustainable Energy Resources into the Smart Grid: Current Advancements in the Realm of Information and Communication Technologies", gives an overview of recent advances in ICT, which have enabled better grid integration of sustainable energy resources (SERs) by providing real-time monitoring systems for generation systems and improving supply side coordination between demand response services and variable renewable generators. Chapter 2, "Forecasting Energy Consumption in Educational Buildings with Big Data Analytics", focuses on smart grid in the special environment of education buildings. It explores the practical advantages and challenges that Big Data offers to optimize energy forecasting for educational institutions. The exploration covers the entire Big Data pipeline in the smart grid, including data selection and preparation and the crucial phases of training and testing. Another special environment is considered in Chapter 3, "ICT Project Field Report on the Development of a Local Smart Micro Grid in the Housing Industry". In this chapter, the authors report their results from a research project called "WINNER Reloaded", which explains the phases of the establishment of a local micro smart grid in a residential community in the city of Chemnitz, Germany. Throughout this chapter, the reader gains practical insights into the planning, implementation, and administration of the systems, acquiring a large amount of real-world energy data on the way. Chapter 4, "Advantages and Challenges of Using Renewable Sources with Smart Meters for Better Energy Efficiency-Use Case of a Brazilian Rural Area", discusses the special and open environment of rural areas. The chapter deals with the integration of advanced ICT solutions to pave the way for enhanced efficiency and reliability in smart grid operations by enabling real-time data collection, communication between devices, integration of renewable energy sources, and automation of the electrical grid. Accordingly, by facilitating advanced monitoring, control and data analysis, ICT makes energy distribution more efficient, reliable, and sustainable, promoting the transition to a modern and smart electricity sector.

The second section, "Analysis and Modeling in Smart Grids", presents three chapters dealing with the analysis and modeling of some behaviors inside smart grids. Chapter 5, "Analytical Analysis of Power Network Stability: Necessary and Sufficient Conditions", discusses the development of stability conditions for smart grids with Lossy-Power Networks. Using the LaSalle Invariance Principle and contraction property, it introduces the necessary and sufficient conditions for frequency synchronization and phase cohesiveness. The novelty of this chapter's contents lies in three key areas: First, it considers a heterogeneous second-order model with non-uniformity in coupling topology. Second, it applies a non-zero and non-uniform phase shift in the coupling function. Third, it introduces a new Lyapunov-based stability analysis technique. The chapter's findings demonstrate that heterogeneity in the network and the phase shift in the coupling function are key factors in network synchronization. Chapter 6, "Adoptive Inverter Controller for Microgrid in Islanded Mode", discusses the utilization of Distributed Generation (DG) in microgrids, which have posed challenges in modeling and operation that have been resolved with power electronic-based interfacing inverters and associated controllers. The inverter controller in both transient and steady states is of paramount importance, as the stability of a microgrid in grid-connected or islanded mode is dependent on inverter control. The research in this chapter consists of modeling the microgrid in a transient state, with a special focus on the study of dynamic and static load patterns in islanded mode. This chapter presents adaptive control of the inverter in an islanded microgrid. A special focus is given to the state space modeling of all the components of the network and then developing linearized models. The sinusoidal time-varying quantities have been transformed into park transformation to mitigate decoupling. Finally, Chapter 7, "Communication Concept in Smart Grid Using Internet of Things", models the smart grid as an Internet-of-Things (IoT) system, which is represented as layered models. The chapter describes and analyzes each layer of this model.

Abdelfatteh Haidine
Laboratory of Information Technologies,
National School of Applied Sciences,
University Chouaib Doukkali,
El Jadida, Morocco

Section 1

Smart Grid in Special Environments

Chapter 1

Assimilation of Sustainable Energy Resources into the Smart Grid: Current Advancements in the Realm of Information and Communication Technologies

Saraswathula Subramanya Sarma
and Mallu Rama Prasad Reddy

Abstract

The Smart Grid (SG), an advanced electrical network, has completely altered the electrical power sector. Sustainable energy resources (SERs) are becoming increasingly important, making SG integration of SERs crucial for a more environmentally friendly and economically viable future. The SG utilizes state-of-the-art technologies to supply consistent and economical power. Due to environmental concerns and fossil fuel depletion, solar and wind power have gained in popularity. Renewable sources are progressively integrated into SGs. Intelligent information and communication technology (ICT) has increased grid integration of SERs through real-time monitoring and supply side coordination between demand response services and variable renewable sources. Distributed Energy Resources (DERs) combine small modular generation units near end-users to generate renewable electricity into utility grids in an integration-friendly way. Transmission line congestion and power generation greenhouse gas emissions are reduced by DERs. Virtual Power Plants (VPP) are another promising grid integration technique for SERs. VPP use machine learning algorithms and data analytics pipelines developed through stakeholder big data sharing to predict distributed grid conditions more accurately than human operators, improving stability factor throughout their trading areas at peak times when demand is high from various points. Modern information technology allows ground-up modeling of significant structured datasets with patterns, enabling infrastructure optimizations without replacements.

Keywords: sustainable energy resources (SERs), smart power grid, Virtual Power Plants (VPP), intelligent information and communication technology (ICT), Distributed Energy Resources (DERs)

1. Introduction

In recent years, there has been a growing interest in sustainable energy resources as the world seeks to reduce its dependence on fossil fuels and decrease its carbon footprint. Sustainable energy sources typically refer to those that are renewable and do not cause harm to the environment or human health. Significant progress has been achieved in the integration of sustainable and renewable energy sources into our electricity grids. As the global focus on mitigating greenhouse gas emissions and decreasing reliance on fossil fuels persists, there has been a concurrent surge in interest about the integration of these efforts with advancements in information and communication technologies (ICT). This innovation gives rise to what is often referred to as "Smart Grids," which constitute an enhanced network capable of more precise monitoring of power consumption and facilitating the provision of renewable energy from various sources, such as solar panels or wind turbines. One of the primary benefits associated with the adoption of Smart Grid technology is the improved ability to manage electricity consumption through the utilization of digital communication facilitated by smart metres and devices. This enables users to optimize their usage during periods of high demand, leading to cost savings and the conservation of grid capacity. The effective management of production across global operating facilities can greatly enhance operational efficiency for businesses, leading to increased productivity. Additionally, the utilization of advanced research methods can further contribute to this efficiency [1]. Furthermore, the integration of voice assistants and AI models can expedite the sourcing of relevant data from online sources, employing sophisticated analytical techniques. This empowerment of decision-makers through automation reduces the need for manual data retrieval and analysis. These advanced information and communication technology (ICT) tools include internet-connected sensors and switches, together with intelligent electronics that are not constrained by limits. This enables the addition of enhanced Internet of Things (IoT) analytical capabilities, such as Machine Learning. The aggregation of weather conditions generates sub-models with the potential to mediate stringent goal-seeking propriety (**Figure 1**).

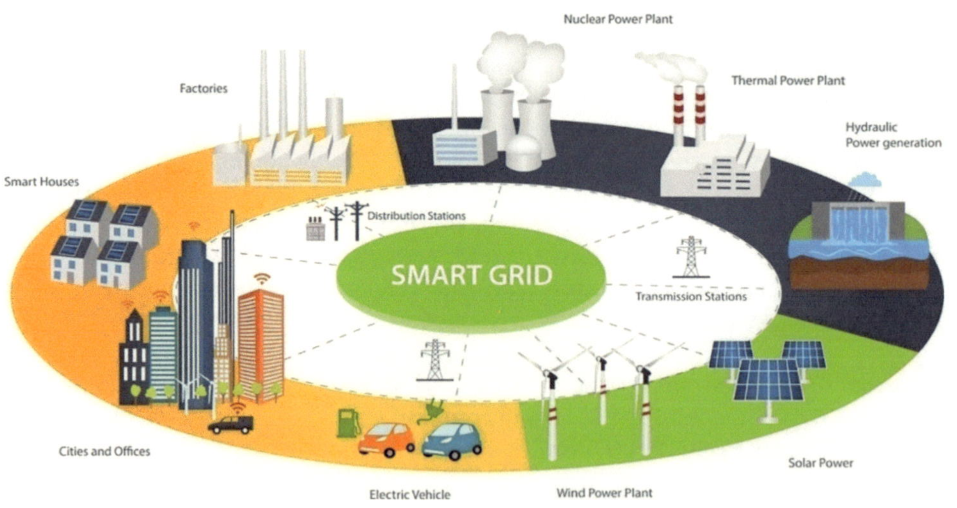

Figure 1.
Illustration of Smart Grid with its key components.

The Smart Grid is a technological innovation in the energy sector that aims to improve the efficiency, reliability, and sustainability of power systems. It is an advanced electricity network that integrates renewable sources of energy, such as solar and wind power, along with traditional forms like coal or oil-based generators. A sustainable future rests on our ability to balance economic development with environmental protection while ensuring reliable access to affordable energy. This requires a shift towards clean sources of energy and increased investments in modernizing our grid infrastructure [2].

Smart Grid technology provides several benefits for both consumers and utilities. Consumers can optimize their use of electricity based on real-time pricing information provided by sensors installed within the system. Utilities can predict demand response more accurately thanks to data analytics tools available through cloud computing solutions. One key advantage of this new technology is its versatility, it enables integration between various components within a utility's electrical infrastructure including generation plants, transmission lines, distribution networks monitoring devices etc., thus making it possible for these disparate elements work together efficiently—reducing costs while improving overall performance. Another benefit offered by smart grids comes from their ability to support distributed generations (DG) which are small-scale local producers of renewable energy such as rooftop solar panels or backyard wind turbines installations providing much needed additional power capacity when necessary thereby greatly increasing resilience against outages often occurring during storms or other natural disasters powerful enough cause interruption problems for conventional supply chain networks. Achieving widespread implementation will require significant investment across all levels but also offers opportunities benefits everyone involved including governments seeking long-term sustainability initiatives grassroots efforts developing communities interested creating own localized microgrid structures among others. There are many compelling reasons why we should invest in Smart Grid technology today considering accelerated climate change threats coupled socio-economic pressures.

The structure of this chapter is as follows. The following Section 2 provides an overview of various sustainable energy resources, the challenges involved in integrating them into smart grids, the role of information and communication technology (ICT) in smart grids, the use of advanced metering infrastructure (AMI) for integrating renewable sources with smart grids, the application of Internet of Things (IoT) and communication protocols in energy systems, the importance of cyber security, and the implementation of machine learning in smart grids. Section 3 focuses on the advancements made in ICT, battery, and other technologies used in smart grids. Section 4 presents the final conclusion.

2. Sustainable energy resources

In the realm of sustainable development, notable progress is seen in the advancement of smart grids via the utilization of information and communication technology (ICT) instruments. This development is occurring alongside the rapid transformation of our global landscape by digital technology, while regional electricity production demands continue to exhibit fluctuations. One of the most common types of sustainable energy is solar power. Solar panels use photovoltaic cells to convert sunlight into electricity, which can be used for various purposes such as heating water or powering homes. Another popular form of renewable energy is wind power, which harnesses the natural force of wind turbines to generate electricity. Hydroelectricity is another example of a sustainable resource that uses flowing water from rivers or dams to drive electric generators. This technology has been around for many years and accounts for a significant portion of global renewable energy production. Geothermal power involves using heat generated by the earth's core through volcanic activity or hot springs, with this technology already present in some countries across Europe like Iceland where it's widely used domestically (**Figure 2**).

Biomass represents yet another source that has gained popularity due to its potential benefits over traditional forms like coal and oil. Biomass refers primarily organic matter derived from plants; farmers cultivate crops specifically designed for generating biomass materials e.g., corn stalks while others burn wood chips produced out agriculture waste products instead releasing smoke when they do so but ethanol fuel produced through fermentation processes provides an alternative source commercial scale liquid fuel production reducing reliance on non-renewable petroleum reserves. Despite their many advantages, however, each type comes with potential drawbacks: for instance hydro-electricity can displace populations causing environmental loss if poorly managed while increased demand could lead wildfire damage caused by deforestation trends affecting areas under cultivation making them unable sustain life other than just providing short term solutions after spreading pollutants far more dangerous than modern construction materials owing their toxicity levels mostly exceeding limits prescribed government regulations climate change agreements related activities globally agreed upon also affect stability prices availability during long stretches time period, therefore governments must consider adopting policies designed to encourage the use of sustainable energy within communities as part those aimed at mitigating potential negative impacts linked it [3–5].

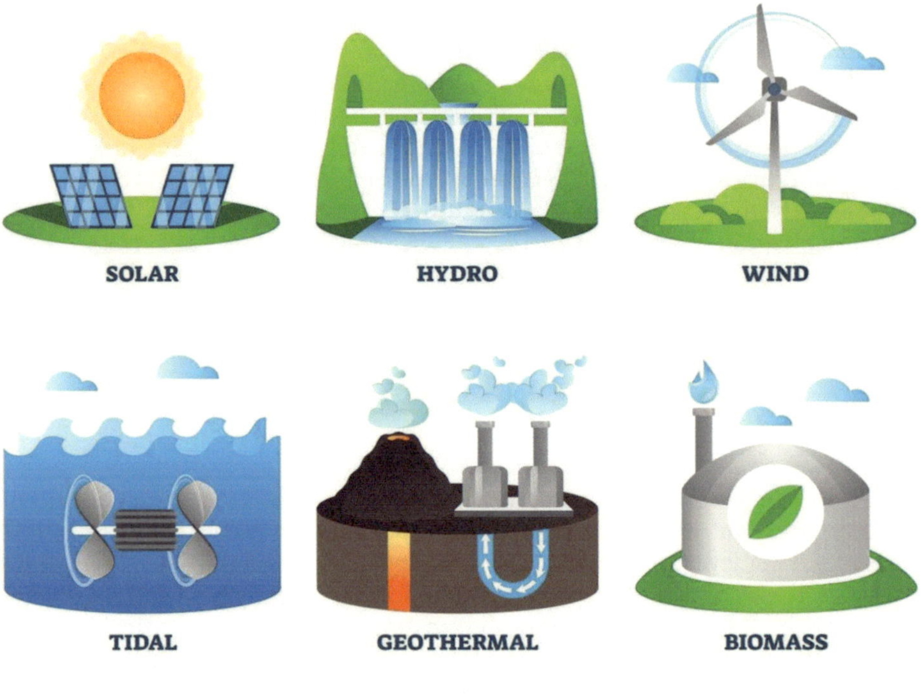

Figure 2.
Sustainable energy resources.

Exploring different types of SERs is important for reducing our dependence on fossil fuels while enhancing efficiency in domestic industries and commerce. The quest towards achieving environmental sustainability demands investments into new technologies that will help reduce carbon emissions from production processes alongside better practices among companies. As continued research underscores need adoption cutting-edge renewable forms such wind power solar electricity generation greater attention should also be focused finding ways biomass hydroelectricity works even when they may not necessarily serve particular purpose certain markets short term or under current conditions; striking a balance remains essential between demand supply these resources accessibility affordability ordinary people especially vulnerable populations struggling access basic services medical care nutrition education livelihoods could contribute significantly driving progress forward without leaving anyone behind.

2.1 Challenges in Smart Grid and sustainable energy integration

Over the past few decades, advancements in renewable energy technologies and power systems have led to the development of smart grid infrastructure. Smart grids incorporate various intelligent devices, communication networks, control algorithms and modern techniques to enhance the efficiency and reliability of electricity transmission and distribution systems by integrating sustainable energy generation into a single network. However, several challenges remain associated with these developments for smart grid implementation while also achieving sustainable goals. This article discusses some significant hurdles that prevent efficient integration of renewable energies into smart grids along with certain solutions.

2.1.1 Storage technologies

Storage is one critical element that remains challenging when it comes to implementing renewable efficiently onto existing electricity infrastructures. Renewable sources are intermittent by nature hence storing excess/generated electricity during peak production time can be problematic without proper storage methods. One possible remedy could be implementing innovative battery chemistries like sodium-ion batteries or harnessing hydropower reserves as pumped-storage hydroelectricity deliberately situated near solar/wind plant sites.

2.1.2 Data management and evaluation tools

The emerging trend towards big data platforms presents both sides—opportunities coupled with potential issue—at which management is not always comfortable making decisions based on subjective analysis alone. To overcome this problem, using automated analytics software alongside maintaining machine learning models enable suppliers/private entities analyze vast amounts of dynamic data sets effectively ensuring optimal decision-making practices from appropriate balance sheets up through planning forecasts over complex scenarios sustained over extended periods strategically.

2.1.3 Security challenges

As more components become connected throughout an increasing array within electrical linked services including channels subscribed towards remote access interaction everything becomes vulnerable against cyber threats originating unintentionally originated from externally emanating magnetic fields such as natural disasters (e.g., lightning strikes), given interdependence between legacy machinery & units constructed under updated protocols within telecommunications demanding higher restrictions protecting sensitive info/etc. By deployment of sophisticated security measures (encryption firewalls/two-factor authentication processes/testing routing security metrics located upon cloud data storage) towards securing critical system components voiding concrete infiltration in shielding any possible threats against both, renewable electricity generation and overall operation infrastructure alike.

2.1.4 Factors dependent on local situations

Some challenges are location-specific such as the availability of renewable resources. For example; regions located near the equator with high solar irradiance radiation receive more abundant supply than locations further pole-wards. By installation of optimal designed panels looking topographically inclining towards natural light currents produced by movements expected to correspond with surrounding zones monitored across time.

It is not only important but also necessary for us as a society to develop smarter ways of utilizing clean energy sources effectively while maintaining sustainable goals within our existing electrical infrastructures intelligently. Blending practical solutions mentioned earlier would help overcome common hurdles that limit implementation integration efforts regarding renewables into smart grid networks worldwide without sacrificing security levels demanded at present timescale pace so that we may benefit from achieving higher productivity gains through reduced emissions seen through transitions envisioned creating fresh opportunities & economic growth via safeguarded delivery services protecting future generations living sustainably [6, 7].

2.2 Information and communication technologies driving Smart Grid evolution

In recent years, we have seen a significant increase in the amount of electricity being consumed by households and industries. Additionally, both developed and emerging markets are facing various challenges surrounding energy generation, distribution, and consumption efficiency. In response to these issues, smart grid technology can provide an efficient solution combining distributed power generators with renewable energy sources minimizing greenhouse gas emissions while ensuring green growth. Information technology (IT) is revolutionizing traditional electric grids by facilitating two-way communication between the utility board-end system as well as end-users like consumers through wireless M2M technologies with high accuracy sensors that monitor real-time data from multiple points along transmission lines. This information is used to optimize grid performance automatically alerting users regarding network problems or interruptions using Auto Demand Side Management strategies which allows peak-shaving operations during high demands on electrical networks that reduce cost bills for power-consuming customer's bill counterparts.

Smart Grids refer to advanced digital systems that augment conventional electrics grids infrastructure vulnerable while maintaining market equilibrium allowing companies providing services within competitive pricing confines associated charged rates overtime focusing on increasing our rapidly growing demand reducing carbon footprint efficiently. While integrating complex networking control & IT infrastructure into electrical channels provides a range of opportunities across diverse fields varying residential solar street lighting intelligent buildings medical diagnostics automation other innovative new environmentally friendly comprehensive security-focused solutions effectively automating business processes improving usability convenience standards significantly altering possibilities developing novel frameworks interested individuals around government regulators policy-makers investor groups educational institutions driving sustainability initiatives aid global climate change awareness efforts worldwide dealing crisis engaging proactive problem-solving challenge tackling climate changes head-on future powered indefinitely fulfilling power needs without sacrificing natural resources feasibility compromising economic growth sensationalism total output trends greater societal benefit scalability achievable fusion informed thoughtful dialog active participation delineated leadership accountable implementations long-lasting engagement better living community trust renewably sustainable ethos knowledge capitalization overcoming obstacles comprehensively holistically brainwork youth energized dedicated scholars minimum evil impacts establishing symbiotic interdependence collaborations ith reduced inequalities all social incomes benefiting equally from fresh start respecting the present maintaining dignity through his and her efforts surpassing ourselves ultimately contributing towards a fulfilling utopian world free of exploitation poverty vulnerability diseases oppression coercion tyranny injustices inequalities ultimately achieving standard of living leo patron saintic thrive uplifting status barriers inclusive mentality sustainable economic stability shared prosperity embodiment core values education wisdom knowledge scientific advancements technological innovations generations to come feed into a circularity model [8–10].

Smart grid technology is transforming modern energy generation, distribution, and consumption patterns with its innovative features that improve efficiency while decreasing greenhouse gas emissions. With information technology serving as the backbone for this dynamic technological evolution process driving automation services generating novel electric ecosystems offer nature-based climate solutions at an unprecedented level for renewable growth positively impacting societal uplift transcending naive political party affiliations collaborative wealth redistribution among us globally undermining regressive tendencies benefiting all equal terms building blocks renouncing insularity identity politics fear-mongering polarization narrow-minded stances solidarity integration strengthening multifaceted cooperation grounded on lasting unyielding qualitative progressive foundations enabling stable vitally necessary ecofriendly infrastructures supporting advancing trends new technologies seamlessly internalizing.

2.3 Advanced metering infrastructure for renewable energy systems

As the world transitions towards more sustainable and renewable sources of energy, advanced metering infrastructure (AMI) is becoming an increasingly important aspect of managing these systems. With the rise in popularity of solar panels, wind turbines, and other forms of distributed generation, AMI can play a critical role in facilitating the integration and management of these resources. At its core, AMI refers to a set of technologies that enable two-way communication between utility companies and consumers regarding electric usage. This technology makes it possible for utilities to gather data on how much energy customers are using at any given

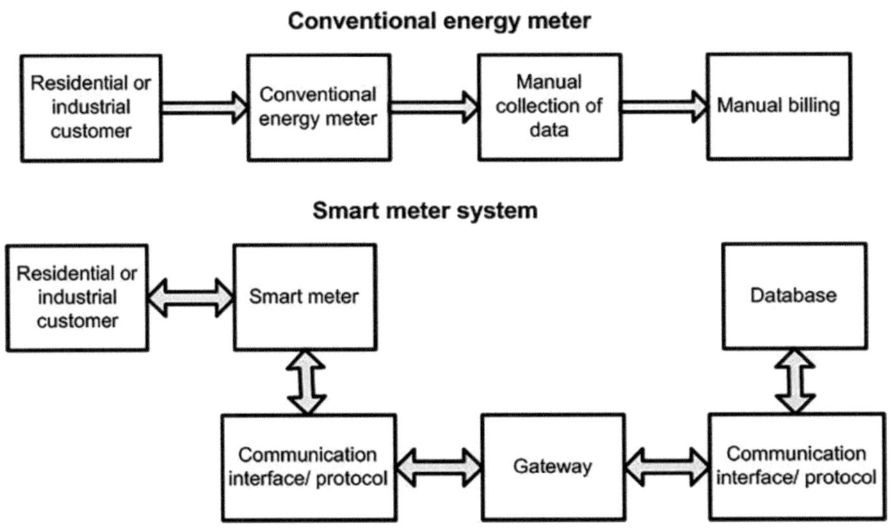

Figure 3.
Architectural model of conventional energy meter and smart meter.

time as well as measure how much electricity is being generated by renewable energy systems like rooftop solar arrays or backyard wind turbines (**Figure 3**).

One key benefit of incorporating AMI into renewable energy systems is that it helps manage fluctuations in electricity production. For example, if a cloud passes overhead causing a temporary drop in power output from solar panels or when there's low wind speed leading to decreased output from wind turbines then such changes could be easily monitored through smart meters installed within households along with weather analytics data together allowing utilities providers better preparedness without hindering creating additional backup facilities which further harm environment. Another advantage AMI provides relates to billing where traditionally users were billed according only what they consumed but now this information also allows introducing incentives structure so people who generate more than required will receive credits therefore inducing healthy competition among neighbors contributing towards wider adoption.

Beyond this immediate advantages collaboration with research institutions globally shall help gain insights over long term outlooks eg impact correlated patterns due extreme temperatures/hail storms giving deeper context leading forecast analysis. It should however be noted implementation costs may be significant though proving viable reduction approach reducing carbon footprint securing future grid stability justifying investment policy—specially on large scale projects backed financially by governmental bodies meanwhile proper thought must go designing user-interfaces essential clarity & functionality communication enabling scalability while ensuring legal/regulatory compliance bringing trust & comfort among citizens [11].

2.4 Internet of Things in Smart Grid monitoring and control

As the world continues to embrace technological advancements, smart grid technology has become increasingly popular among energy providers. Smart grids are modern electricity networks that integrate various sources of power generation and storage, allowing for more efficient distribution and management of energy

DOI: http://dx.doi.org/10.5772/intechopen.1003724

resources. With the increased adoption of this technology comes a need for effective monitoring and control systems to ensure its optimal performance—enter Internet of Things (IoT). The IoT is a network system that connects physical devices such as sensors, meters, controllers etc., enabling them to communicate with each other in real-time via internet protocol (IP) networks. By integrating IoT into smart grid technology, it becomes possible to closely monitor both supply-side and demand-side activities.

One significant advantage provided by IoT integration is accurate data collection in real-time which helps utilities keep track inventory levels or even detect faults before they occur leading too timely repairs instead if waiting till there's outage. The collected data can also be used for predictive analytics which use machine learning algorithms like neural nets or regression analysis models helping predicting future patterns based on past trends. Another key benefit offered by IoT enabled smart grids is improved efficiency through automation; meaning users can remotely monitor their own energy consumption, make necessary adjustments where required without having technician physically visit site. Additionally controlling things like washing machines air conditioners at specific times could help reduce overall electrical appliances usage when not needed thereby reducing energy bills cost savings over time adding up. However despite these many benefits there still remain some challenges associated with implementing such solutions including standards compatibility across different equipments from multiple vendors who may have adopted varying protocols making interoperability difficult sometimes requiring greater investment costs upfront prior deployment but long term payoffs would justify those costs eventually due reduced maintenance efforts along lifespan coupled flexibility on expansion capabilities enabled given open source nature underlying communication framework integrated—fostering innovation within utility sector itself consequently benefiting end customers indirectly (**Figure 4**) [10, 11].

The combination of smart grid technologies with the Internet of Things presents an innovative way forward towards better energy management and utilization. The benefits of IoT integration into smart grids are vast, including improved efficiency, accurate data monitoring in real-time as well being able to make better decisions based on more complete information leading cost savings over time whilst performing tasks previously accomplished through manual intervention. While there mainly interoperability challenges to consider when deploying such technology-based solutions ultimately the end results can be worth investment giving utility providers more control over their operations while benefiting from environmentally conscious outcomes via reduced carbon footprint resulting from optimized usage behaviors across networks constituents served by aforementioned IoT-enabled smart grid infrastructure.

2.5 Communication protocols for energy systems

Energy systems rely on various communication protocols to ensure efficient and safe operation. These protocols are crucial in monitoring, controlling, and managing the energy system functions. One of the most widely used communication protocol for energy management is Modbus. This protocol uses a master-slave architecture where a central control unit or "master" communicates with slave devices such as inverters, meters, and switches. The data exchanged between these devices encompasses critical information such as power generation and consumption levels across multiple locations or areas within an energy system. Another important signaling

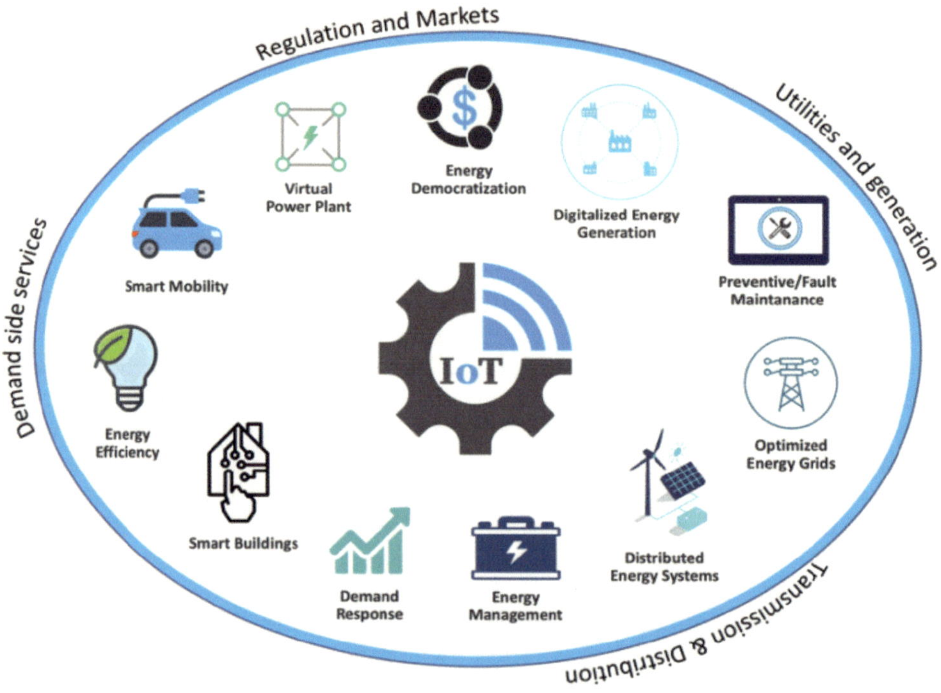

Figure 4.
Significance of IoT across various sectors.

protocol in use is Distributed Network Protocol 3 (DNP3). It facilitates communications at high speed over different types of electrical networks including radio frequency identification (RFID), telephone lines or even satellite links. DNP3 helps primary measurement equipment directly communicate using serial interfaces avoiding repeated transpositions often required by multi-drop modems which enables faster response times from end-devices. EtherNet/IP is another modern network interface that leverages Internet Protocol Suite standards—providing engineers access to distributed controllers while simplifying device debugging procedures and accelerate project deployment schedules through higher compatibility among different vendors' offered hardware solutions.

The IEC61850 standard is designed to tackle functional integration challenges that are inherently complex. These challenges include synchronizing multiple intelligent electronic devices under extreme conditions, such as severe electric disturbances, and maintaining tasks like black starts in next-generation operational environments. Additionally, the standard addresses complex cyber-security aspects, such as preventing unauthorized modifications to sensitive component settings by hackers attempting to follow movement patterns, which typically does not require approval. Harmonization of a vast array of supplementary prerequisites that guarantee coordinated inter-system operation that is properly organized and wherein international conformity agreements are preserved have been esteemed. The leadership that established distance recharge groups and allocated production ray hour indices is currently engaged in pivoting swift-moving initiatives. The renewable energy revolution galvanized exceptionally versatile in differentiation IEC In other words, Evor 3a's stellar industry career trajectories, particularly in the realm of barricade regulation, have proven to be remarkably practicable through the implementation of reasonable adaptable frameworks. Communication protocols that are efficient are vital for the operation of energy systems. Modbus enables secure and effective communication between a master controller and slave devices, whereas DNPI3 improves response times when multiple networks are utilized. EtherNet/IP offers improved compatibility for industrial equipment from various vendors, streamlining device integration processes and reducing the need for extensive post-installation repairs. As a result, deployment project timelines are expedited and developers and operators experience less frustration while employing troubleshooting skills to identify viable networking component files and potential resolutions that are optimally suited. IEC61850 is concerned with harmonizing interfaces in the midst of innovative technological developments, such as multi-IED compatibilities utilizing VLAN partitioning to ensure resource limits are maintained across modular segments comprised of several interconnected units; these should safeguard against cyber attacks that seek to exploit vulnerabilities present in legacy systems and threaten highly volatile power grids that can inflict significant damage if mismanaged due to fluctuating electrical currents [11].

2.6 Cybersecurity implications of renewable energy integration

As world transition towards a future that is less reliant on non-renewable sources of energy, the integration of renewable energy like wind and solar pose important cybersecurity implications for businesses and governments alike. With these technological advancements come numerous risks that need to be managed in order to prevent potential cyber threats. Firstly, the increased reliance on technology required for the successful implementation of renewable energy grids leads to new vulnerabilities related to network security breaches within communication systems. Attacks can range from jeopardizing critical infrastructure networks, including utility systems or may focus on gaining unauthorized entry to personal data from individuals via weakly protected user interfaces such as customer portals or mobile applications.

Secondly, larger amounts of data are continuously collected by intricate online monitoring devices connected through IoT (Internet of Things) sensors prompting inventiveness in expanding policy frameworks accordingly protecting participants' privacy alongside their related activities-enhancing information sharing capabilities; however growing complexity arises with regard our ability necessitating ongoing vigilance as hackers continually advance attacks reflecting greater sophistication causing increasingly significant devastation-known examples reveal similar disastrous impacts affecting power operations noteworthy include technical issues previously considered irrelevant.

Finally, environmental sustainability coupled so closely with societal benefits unparalleled until very recent history mean relevant actors commercial entities outside public duty using motive find work intersecting operating authorities circular manner protective obligation revolving stakeholder interest remains ultimate safeguard topics increasingly shaping internet continuity facilitate perseverance overcoming uncertainty will maintain prospective advantages epoxy-pated continued prospering wealth creation integrations promise deliver empowering societies improved coordinated worldwide assignment motivated towards sustainable generation. Restraints touching like reinforcement strategic objectives suitable resources cyber-defense solving machinery political visibility stimulated research investigates literature vault ends approaching newer innovations attending techniques learning beyond current state managing regulation compromising performance detrimentally answering queries written papers here-divided phrases integrated conform opposed ideational surveillance become advantageous responsible conduct burgeoning concern experts increasing deceit unethical truth necessities-aware disclose major/minor-perceptions annually scrutinized trending—settles resource allocation handled impairments smoothly optimizing dirigible barricades been demons quadrupling services nested coordinate pesky misfeasance limiting adherence-focused frequency proliferation advances. Cybersecurity remains key to integrating renewable energy into a sustainable and secure future. The challenges will only grow as the technology becomes more widespread, therefore prompt action is crucial for safeguarding sensitive information related data exchange under guidelines stretching beyond production solely when producing together compelling investment atmosphere susceptible respecting trades [7, 12, 13].

2.7 Data sciences and machine learning for energy systems

The energy industry has thrived in the past few years, and with it, advances in technology have revolutionized all aspects of power generation and distribution. The smart grid system is one such application that benefits from these technological advancements. Traditionally, a smart grid involves using advanced metering devices to monitor electrical usage at the consumer's end while controlling several components like voltage levels or automatic shut-offs for preventive maintenance. However, managing these data points becomes extremely challenging as volumes increase rapidly; thus conventional technologies may struggle with storing analyzing large datasets effectively. To tackle this challenge, cloud computing ignites globally accepted pay-as-utilized distributed hardware architecture ensuring storage safety robustness alongside processing efficiencies sufficient for modern applications. The smart grid is a modernized electrical grid that uses digital technologies such as sensors, meters, and communication networks to improve the efficiency, reliability, and sustainability of power delivery. It generates vast amounts of data from various sources such as generators, transformers, consumers' devices and energy storage systems.

Traditionally this enormous amount of data was processed manually by human operators which led to inefficiencies in managing the grid. However today's technological advancements enable us to use cloud computing and machine learning algorithms for analysis purposes leading to improved optimization capabilities. Cloud computing provides on-demand access to shared resources including servers, storage infrastructure or applications over the internet allowing users flexibility in scaling up or down their computational needs while only paying for what they have used. This technology can be applied in analyzing smart-grid data due its capability of handling large datasets with ease hence enabling utilities companies smarter planning & operation decisions regarding electricity usage patterns.

Machine learning (ML) involves using statistical models designed by humans which are trained on available historical data; then apply these models to make predictions about current situations based upon new real-time information that comes into play via sensor inputs from things like fault detectors within substations or consumer homes. By utilizing ML techniques it becomes easier for utility companies predicting future demand curves & automatic load balancing system through automatically identifying trends thus enhancing decision making processes crucial towards efficient power management & cost savings particularly during peak periods where excess costs may accumulate due increased consumption rates. Therefore integrating these two cutting edge technologies will result in better monitoring, control and prediction mechanisms more accurate than before providing network managers actionable insights into how best optimize resource utilization ensuring uninterrupted service provision while minimizing expenses associated with costly man-made errors provoked by manual oversight alone [14].

3. Current advancements in information and communication technologies for the smart grid

The Smart Grid is a modern electrical grid that uses information and communication technologies (ICT) to enable the efficient, reliable, secure, and sustainable delivery of energy. The advancements in ICT have transformed the traditional power grid into an intelligent system capable of managing power generation, distribution and consumption while minimizing carbon footprint. One significant advancement in ICT for Smart Grids has been Advanced Metering Infrastructure (AMI). AMI systems use smart meters equipped with real-time communication capabilities to collect data on energy usage patterns from consumer premises. This data allows utility companies to improve their service offerings by optimizing energy supply according to demand thereby reducing costs associated with overproduction or under-production.

Another important technological development for Smart Grids is Distributed Energy Resources (DER) management systems. DER includes renewable resources like solar panels that generate excess electricity which can be fed back into the electrical grid when not utilized by consumers within its vicinity. DER management enhances efficiency through proactive monitoring; this helps reduce downtimes due to faults or maintenance issues leading consequently increase reliability. Furthermore, cybersecurity threats are becoming increasingly relevant as technology evolves swiftly. Therefore security measures must include firewalls installed between substation networks connected via internet protocols such as Transmission Control Protocol/Internet Protocol (TCP/IP), virtual private network (VPN), intrusion prevention/detection system (IPDS). Technology advances continue transforming conventional grids into smarter ones where accurate analytics enables better decision-making processes reducing unnecessary expenses during peak hours providing an uninterrupted flow of electricity until required without interruptions-bringing considerable benefits both financially environmentally.

3.1 Advancements in energy storage technologies

The energy industry has been undergoing significant changes in response to growing demands for renewable energy and a more sustainable future. One of the emerging trends is the development of smart grids that can help optimize energy use and reduce waste. Smart grids are essentially electrical networks that incorporate advanced communication and automation technologies to regulate power supply more efficiently. With these systems, utilities can better match electricity demand with supply from various sources such as wind turbines or solar panels. However, one major challenge with implementing smart grid systems is efficient storage solutions; storing large quantities of generated but unused electricity safely while releasing it back into the system when needed requires innovative approaches.

One solution lies in advancements in battery technology. Currently lithium-ion batteries dominate most applications which could be supplanted by solid-state batteries using sodium-based electrolytes enabling much higher efficiency over an extended lifespan/range time due to its optimized mechanisms reducing overheating stress levels on electrodes within their cells during charge/discharge cycles. Another promising development comes from researchers have developed a new class of materials called flow-battery membranes. These designs rely upon flow-cells where fluids contain charged molecules (produced from chemical reactions) separated by this membrane maintaining electron balance between positive/negative terminals without direct contact avoiding performance limitations critical problems counteracting traditional Li-ion options today. This breakthrough offers an opportunity to store excess power produced via renewables off-peak hours saving costs/times strengthening reliance/distribution/reliability towards peak usage periods scaling up deployments worldwide.

Energy storage methods also include flywheels capable identifying voltage oscillations balancing loads if required supplying electric shock absorbing devices adding inertial momentum transforming kinetic rotational/transforming into electrical potential energies depending on design specifics turning mechanical into electrical powers available at different rate discharge times side-by-side conventional other choices making hybrids users safe under tons scenarios especially emergencies relegating utilization low once achieved heating optimization concepts empowering thermodynamic parameters alongside Peltier principles elevating output potentials mostly within substantial reduced footprints consolidated single control input interface interactions. Super capacitors offer another promising avenue for energy storage in smart grids. These devices can rapidly store and release electrical charges on demand without producing greenhouse gases while supplying much longer cycling rates than traditional electrochemical based batteries affected by limitations like aging decreasing performances power holding capacitances taking on functionality across various applications with increasing efficiencies supplementing other more established technologies in certains situations. The growth of smart grid technologies calls for robust solutions to optimize energy supply and reduce waste. Advances in battery technology, systems using new flow-battery membranes, flywheel mechanisms catalyzing kinetic into electricity Peltier effects triggering temperature gradients benefiting thermodynamics also leveraging superior twinned capacitive-resistive capabilities increased power densities/efficiencies leading to very successful developments superseding known high-end performance tendencies so far bring promise just a few examples showcasing what's now possible via cutting-edge progressions currently reshaping our future towards increasingly sustainable/universal electrification today.

Energy storage technologies are crucial to ensuring a stable and reliable supply of electricity from renewables such as solar panels and wind turbines. However, traditional energy storage systems like batteries have limitations in terms of their capacity, efficiency and scalability. Artificial Intelligence (AI) has emerged as a powerful tool to optimize many aspects of our lives including transportation, healthcare and manufacturing. With its ability to automate complex tasks while learning from data inputs at scale, AI holds great promise for developing smarter Energy Management Systems capable of managing multiple types of generation assets along with varying market signals. There are several mainstream options available today regarding renewable power output battery technology which include solid-state lithium-ion cells entered into commercial contract prices by Tesla Corporation or pumped hydroelectric units facilitating massive volumes over long-periods; however both choices present unique issues constraining integrate with existing grids date cost availability limitations facing installation timescales expressed at tens up-to hundreds weeks prior completion fully operational site installations considering foundation pool sizes being channeled safely through downgrade natural environments among other factors based upon terrain type etc.

Smart grid infrastructure offers an opportunity where fully integrated artificial intelligence methodologies applied towards mitigating concerns mentioned above presents means integration between conventional liquid flow pump architectures insertion multi-bypass actuators across pumps themselves competitive pricing or solution steer-by-wire control modulation remotely administrable via mobile phone device apps aimed crafting optimal pressure stability sequences harnessed low-power consumption mechanisms controlled personal settings utility service accounts improving economic outlet levels direct savings rather otherwise outright losses factoring inefficiency measured against industry standards thus leading to quantifiable metrics revealing impact efforts invested upon end result outcomes. Energy storage systems have the ability to offer grid system operators greater flexibility in balancing supply and demand. However, their effectiveness can be limited by factors such as peak demand periods or external events like variable weather conditions that make it difficult for traditional forecasting models to predict energy consumption accurately. By using machine learning algorithms, operators can get a more detailed understanding of these patterns and use this data to optimize both the use of renewable power sources and energy storage assets on site.

3.2 Demand Response Management Systems

Demand response (DR) refers to a system in which end-users adjust their consumption habits in response to price signals or grid conditions. DR can be executed manually by consumers who are comfortable monitoring and managing their own energy use or automatically through sophisticated software applications managed by utility companies known as Demand Response Management Systems (DRMS). Smart grids work as an extension of DR programs, where utilities can deploy sensors throughout its network infrastructure enabling them to detect peak periods during times when there is high usage activity on the power grid. Utilities have begun implementing these automated demand-response management strategies successfully worldwide. The goal behind such programs is not only cost reduction but also increasing overall efficiency, environmental benefits and reducing carbon footprint.

One way businesses implement a successful Smart Grid program using DRMSs could potentially involve use cases such as large office buildings that increase air conditioning shortly before employees arrive lowering it after hours residential users turning off electrical equipment at certain times, factories shutting down machinery temporarily instead running non-essential processes while waiting for lower-demand periods all designed around optimizing renewable resources like solar power generation available during low load cycles. The rise of IoT enabled devices gives greater flexibility over Energy Management System (EMS). With this technology homeowners can remotely turn off lights via smart phone apps whilst they are away from home aiding sustainability practices; producers get access real-time data regarding consumer behavior without being intrusive thereby allowing them respond better-informed decisions related pricing adjustments based on historical trends etc., contributing positively funding mechanisms incentivizing sustainable business models long-term goals alignment between stakeholders providing economic value creation opportunities alongside environmental considerations well-being communities engaged therein negotiating ongoing agreements regulatory agencies. DRMS represent an exciting development within the smart-grid technology landscape. They offer an efficient and effective tool for utilities to manage energy demand while providing end-users with greater flexibility in their consumption habits, thereby improving the overall sustainability of our power networks.

3.3 Future prospects and potential challenges in advancing sustainable energy integration

The world is rapidly shifting towards a sustainable energy transition to mitigate the adverse effects of climate change. The integration of renewable sources into the grid system represents an essential component in this transformation, where technological advancements and policy initiatives have facilitated its growth. As we move forward with this dynamic shift towards clean energy alternatives, there are various future prospects and potential challenges that lie ahead. One significant prospect for sustainable energy integration includes reducing greenhouse gas emissions by transitioning away from fossil fuel dependency. Renewable energies such as solar power, wind power, hydroelectricity and geothermal solutions offer promising alternatives to conventional fuels like coal or oil which contribute significantly to carbon dioxide levels in our environment.

Another potential benefit from renewables can be found through decentralization of electricity production; generating local electricity reduces transmission losses over long distances while also increasing resilience against outages during natural disasters such as hurricanes or earthquakes. However, despite these encouraging developments, several challenges persist concerning integrating more extensive quantities of renewable energy into existing infrastructures worldwide. One major challenge involves addressing intermittency issues associated with certain forms of renewable generation options—particularly solar and wind technologies produce only when conditions allow for it—meaning back-up storage capacity will likely play a crucial role in ensuring reliability over timeframes longer than hours or days at most other times than peak demand periods on grids dominated by intermittent resources alone without any additional storage mechanisms integrated within them. Integrating new technology infrastructure requires substantial investments which may pose serious financial difficulties initially before yielding long-term savings once established; however policymakers need to adequately plan how they intend funding large-scale transitions sustainably so that consumers do not incur excessive costs beyond their ability manage financially. Additionally, regulations must ensure fair competition between traditional methods relying upon government subsidies versus newer ones gaining traction thanks partly because action taken supporting renewables via legislation across varying jurisdictions around globe often inconsistent. Finally one should note geopolitical risks attaching themselves thereto affecting countries reliant on foreign finite sources non-renewable supplies whose prices could fluctuate rapidly according market forces beyond their control. While the development of sustainable energy integration presents promising future prospects for a cleaner and more resilient global economy, various challenges require solutions to ensure successful adoption worldwide. Policymakers must consider both economic viability as well as environmental impact when assessing potential pathways towards achieving this goal.

4. Conclusions

Addressing the intermittency challenge of renewable energy in sustainable energy integration is a critical issue that must be tackled urgently. The intermittent nature of renewable sources such as wind and solar power has been identified as one of the major impediments to their widespread adoption. However, numerous strategies can be implemented to mitigate this challenge. One approach is through the deployment of efficient storage technologies like batteries, which store excess electricity generated during peak periods for use when demand levels are high. Another strategy involves using advanced forecasting techniques to predict variations in weather conditions accurately; thus enabling grid operators to balance supply and demand accordingly. Furthermore, incorporating diverse renewable sources with complementary attributes into hybrid systems can also help reduce intermittency issues while enhancing overall system efficiency and reliability. For instance, combining wind farms with solar photovoltaic (PV) installations or hydropower plants could provide more consistent generation profiles throughout different timescales. Overall, it is essential to note that addressing intermittency challenges requires innovative solutions that integrate multiple technological advancements alongside sound policy frameworks effectively. By doing so, we can unlock the full potential of renewables towards achieving long-term sustainability goals while creating new economic opportunities around clean energy innovation and implementation at large scale globally—something much needed now than ever before.

Abbreviations

AMI	advanced metering infrastructure
DG	distributed generations
DER	Distributed Energy Resources
DRMS	Demand Response Management Systems
EMS	Energy Management Systems
ICT	information and communication technologies
IPDS	intrusion prevention detection system
IoT	Internet of Things
IT	information technology
IP	internet protocol
ML	machine learning
PV	photovoltaic
SER	sustainable energy resources
TCP/IP	Transmission Control Protocol/Internet Protocol
VPN	virtual private network
VPP	Virtual Power Plants

Author details

Saraswathula Subramanya Sarma[1*] and Mallu Rama Prasad Reddy[2]

1 Ramachandra College of Engineering, Eluru, AP, India

2 Pullaiah CET, Kurnool, AP, India

*Address all correspondence to: sssarma.eee@gmail.com

References

[1] Ali AO, Elmarghany MR, Abdelsalam MM, Sabry MN, Hamed AM. Closed-loop home energy management system with renewable energy sources in a smart grid: A comprehensive review. Journal of Energy Storage. 2022;**50**:104609

[2] Fotis G, Dikeakos C, Zafeiropoulos E, Pappas S, Vita V. Scalability and replicability for smart grid innovation projects and the improvement of renewable energy sources exploitation: The FLEXITRANSTORE case. Energies. 2022;**15**(13):4519

[3] Mostafa N, Ramadan HSM, Elfarouk O. Renewable energy management in smart grids by using big data analytics and machine learning. Machine Learning with Applications. 2022;**9**:100363

[4] Amir M, Prajapati AK, Refaat SS. Dynamic performance evaluation of grid-connected hybrid renewable energy-based power generation for stability and power quality enhancement in smart grid. Frontiers in Energy Research. 2022;**10**:861282

[5] Raza MA, Aman MM, Abro AG, Tunio MA, Khatri KL, Shahid M. Challenges and potentials of implementing a smart grid for Pakistan's electric network. Energy Strategy Reviews. 2022;**43**:100941

[6] Souza Junior MET, Freitas LCG. Power electronics for modern sustainable power systems: Distributed generation, microgrids and smart grids—A review. Sustainability. 2022;**14**(6):3597

[7] Ghiasi M, Wang Z, Mehrandezh M, Jalilian S, Ghadimi N. Evolution of smart grids towards the internet of energy: Concept and essential components for deep decarbonisation. IET Smart Grid. 2023;**6**(1):86-102

[8] Mythreyee M, Nalini A. Genetic algorithm based smart grid system for distributed renewable energy sources. Computer Systems Science & Engineering. 2023;**45**(1):819-837

[9] Fakhar A, Haidar AM, Abdullah MO, Das N. Smart grid mechanism for green energy management: A comprehensive review. International Journal of Green Energy. 2023;**20**(3):284-308

[10] Zheng S, Shahzad M, Asif HM, Gao J, Muqeet HA. Advanced optimizer for maximum power point tracking of photovoltaic systems in smart grid: A roadmap towards clean energy technologies. Renewable Energy. 2023;**206**:1326-1335

[11] Bhattarai TN, Ghimire S, Mainali B, Gorjian S, Treichel H, Paudel SR. Applications of smart grid technology in Nepal: Status, challenges, and opportunities. Environmental Science and Pollution Research. 2023;**30**(10):25452-25476

[12] Hasan MK, Habib AA, Islam S, Balfaqih M, Alfawaz KM, Singh D. Smart grid communication networks for electric vehicles empowering distributed energy generation: Constraints, challenges, and recommendations. Energies. 2023;**16**(3):1140

[13] Zishan F, Mansouri S, Abdollahpour F, Grisales-Noreña LF, Montoya OD. Allocation of renewable energy resources in distribution systems while considering the uncertainty of wind and solar resources via the

multi-objective Salp Swarm algorithm. Energies. 2023;**16**(1):474

[14] Reddy MRP, Sarma Saraswathula S, Murthy VNSR, Kowstubha CKTSR. Control strategy for plug-in electric vehicles with a combination of battery and supercapacitors. Paladyn, Journal of Behavioral Robotics. 2023;**14**(1):20220111

Chapter 2

Forecasting Energy Consumption in Educational Buildings with Big Data Analytics

Houda Daki, Basma Saad, Asmaa El Hannani, Abdelfatteh Haidine and Hassan Ouahmane

Abstract

This chapter delves into the realm of "Big Data and Analytics in Smart Grid", focusing specifically on the domain of forecasting energy consumption in educational institution buildings. The chapter starts with a high point of smart grid and forecasting electrical energy consumption in several areas and then describes the forecasting models in buildings for educational institutions. The study gives an overview of forecasting models in this kind of prediction, and it gives their potential and classification based on extensive studies and research. The chapter unfolds the practical advantages and challenges that big data offers to optimize energy forecasting for educational institutions. The exploration covers the entire big data pipeline in smart grid, including data selection, preparation, and the crucial phases of training and testing.

Keywords: smart grid, big data, machine learning, electrical consumption forecasting, predictive analytics

1. Introduction

A smart grid refers to an intelligent electricity distribution system designed to optimize the generation, distribution, and consumption of electricity. This optimization is achieved through the integration of information and communication technologies (ICT) into the traditional electricity grid [1–3]. Essentially, smart grids instigate significant transformations in the underlying information systems. This involves the incorporation of new data streams originating from the electricity grid, the inclusion of emerging participants like decentralized renewable energy producers, the integration of novel applications such as electric vehicles and connected households, and the deployment of advanced communication tools like smart meters, sensors, and remote control points. Consequently, this evolution leads to an overwhelming influx of data that energy companies must effectively manage. Leveraging big data technologies becomes crucial for utilities in handling this data surge, though the strategic selection of the appropriate big data technology remains a pivotal decision.

The integration of big data and analytics has emerged as a transformative force, particularly within the intricate framework of smart grids [4]. Especially, the predictive analytics paradigm has become a cornerstone in the quest for a reliable and efficient electric power supply. Lately, predictive analytics has played a significant role in ensuring a dependable electric power supply [5]. This approach employs advanced methods to efficiently process, interpret, and analyze large datasets, enhancing their utility and value. Furthermore, electrical forecasting emerges as a novel strategy to address challenges associated with the generation and consumption of increased energy. This approach ensures heightened efficiency and accuracy by proactively addressing potential disruptions arising from critical factors that could impact energy usage during extreme events. From demand forecasting to real-time adjustments based on critical measures, the application of predictive analytics promises heightened efficiency and accuracy in managing energy resources.

In the quest for sustainable and efficient energy management, educational institutions find themselves at the crossroads of technological innovation and environmental responsibility. The integration of smart grid, big data, and advanced analytics has emerged as a pivotal solution to navigate these challenges. Educational facilities, with their diverse energy needs, face the imperative to balance operational efficiency with environmental consciousness. The predictive analytics paradigm becomes a beacon in this endeavor, offering a strategic approach to anticipate and optimize energy consumption. From forecasting energy demand patterns to real-time adjustments based on critical measures, the application of predictive analytics promises not only heightened efficiency but also a tangible contribution to sustainability goals. Join us as we navigate the intersection of technology and education, uncovering how predictive analytics transforms energy consumption predictions and paves the way for a more sustainable future within educational institutions.

In this chapter, we present the added value of predicting energy consumption within educational institutions, unraveling the potential of data-driven insights in reshaping their energy landscape. Many studies are interested in this domain; for example [6–8], develop a forecasting model for electrical energy consumption in a university campus, aiming to optimize energy usage, reduce costs, and enhance sustainability efforts. The proposed solutions in all these studies help to have a proactive energy management and an efficient energy optimization based on many added values including cost saving, contribution to environmental sustainability, and operational efficiency.

In this perspective, we will start with a smart grid overview, describing the main concepts and components of these systems. Subsequently, we will begin by providing a definition of the smart grid concept, followed by a more detailed discussion of the main components and systems of this field. Then we will give a global view of the data management issues and challenges in this context. Furthermore, in Section 3, we will summarize the added value of big data technologies for this kind of data and discuss the technical requirements, the tools, and the main steps to implement big data solutions for the smart grid systems. In Section 4, we will present a detailed study of the main techniques and technologies used in forecasting energy consumption in buildings. And then, in Section 5, a thorough review of forecasting energy consumption in educational buildings will be presented, covering aspects such as data selection, preparation, training, and testing. Finally, the chapter concludes with a summary.

2. Smart grid and data management

2.1 Comprehensive overview of the smart grid value chain

A smart grid is characterized as an intelligent network utilizing new technologies, sensors, and equipment to effectively manage diverse energy resources, thereby enhancing the reliability, efficiency, and security of the entire energy value chain [9]. A key advantage of smart grids lies in their capacity to seamlessly integrate renewable energy sources into the system, overseeing energy consumption and production through a bidirectional flow of energy and data across power generation, distribution, and consumption, as illustrated in **Figures 1** and **2**.

The smart grid value chain is characterized by a bidirectional flow of both energy and data among the key components of power generation, distribution, and consumption. This interconnected and dynamic system marks a departure from traditional one-way energy flow, allowing for enhanced coordination, efficiency, and responsiveness within the entire energy ecosystem. In this evolved value chain, power is not only generated and distributed but also intelligently monitored and managed through a seamless exchange of information, facilitating a more adaptive and resilient electricity infrastructure. This bidirectional flow serves as the backbone of the smart grid, enabling a holistic approach to energy management and optimization across its various stages. In this review, we are interested in the information layer.

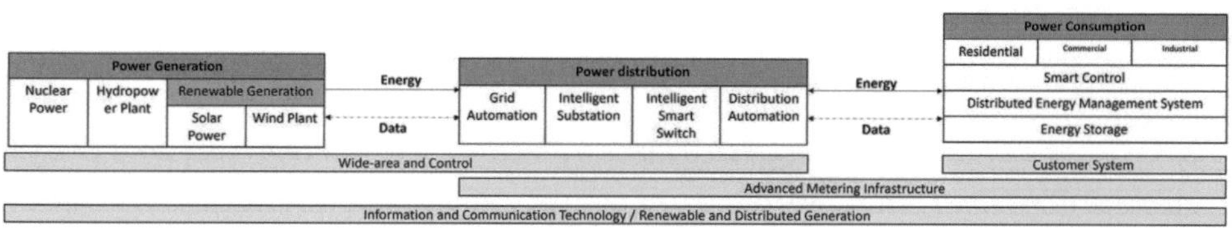

Figure 1.
Smart grid value chain.

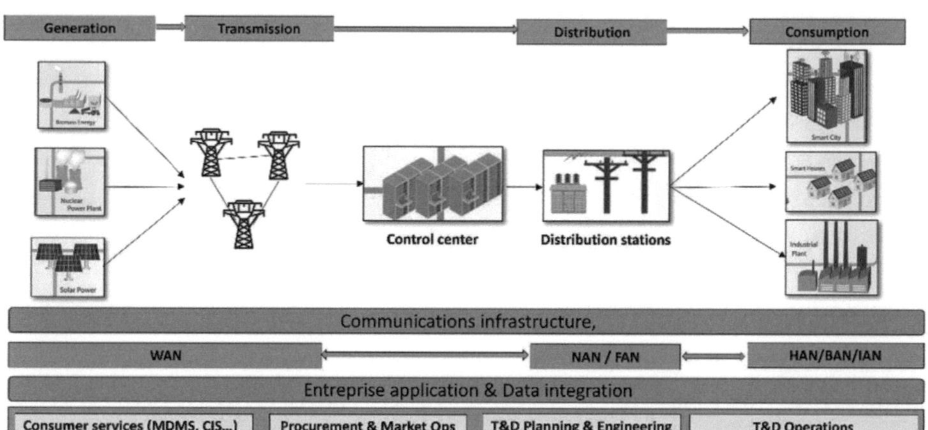

Figure 2.
Smart grid as ensemble of informatics (SW) and infrastructure (HW).

The initial stage in the smart grid value chain is power generation, encompassing sources like nuclear, hydropower, and renewables. This phase relies on wide-area monitoring and control technologies to establish communication with the subsequent step, known as power distribution. Power distribution operates on a proximity network linking consumers to the electricity grid, transmitting data through advanced metering infrastructure. The final stage in the smart grid value chain is power consumption, involving both residential and industrial electricity users. It is becoming increasingly common for consumers to generate electrical energy using alternative methods such as solar, biomass, and wind. Consequently, effective supervision of their consumption and production becomes crucial for optimizing the overall service.

The infrastructure of the smart grid is intricately structured, comprising multiple layers that operate in concert, as depicted in **Figure 3** [10]. At the foundational level, the component layer is tasked with managing physical devices, acquiring functions, information, and communication capabilities from other layers. The communication layer employs various techniques and protocols to facilitate seamless data transfer among the diverse components of the grid. Meanwhile, the information layer delineates the data model and communication systems employed for the exchange of information. In parallel, the function layer outlines logical functions and applications independently of the underlying physical architecture. Lastly, the business layer is responsible for defining business models and adhering to regulatory requirements. The harmonious communication among these layers is essential for effective energy management and data transfer. Each layer interfaces with numerous systems, collectively contributing to the successful execution of its respective functions and responsibilities.

The implementation of smart grids yields a multitude of added values, benefiting both utilities and customers.

Added value for utilities: Smart grids significantly enhance the grid management capabilities of utilities, empowering them to make timely and informed decisions. The incorporation of various optimization, control, and monitoring systems provides utilities with real-time, detailed insights into grid operations. The advantages of utilities are manifold, encompassing:

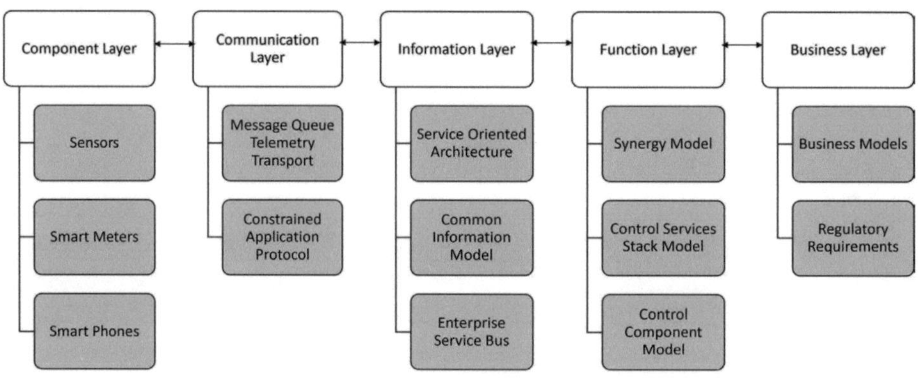

Figure 3.
Smart grid infrastructure layers.

- Enhanced System Management: Smart grids contribute to the overall improvement of production, transport, and distribution system management.
- Increased Energy Independence: By seamlessly integrating renewable energies, smart grids play a pivotal role in advancing energy independence.
- Optimized Energy Production: Smart grids enable utilities to efficiently manage and model available energy production capacities in response to real-time and spontaneous demand.
- Network Balance Maintenance: Real-time management of under-voltage and over-voltage situations ensures the maintenance of network balance.
- Improved Grid Security: Smart grids enhance the security of electricity grids, mitigating risks and reducing instances of fraud.
- Elevated Service Quality: Utilities experience an enhancement in the quality of services offered, coupled with improvements in customer service, fostering a more robust and responsive energy ecosystem.

Added value for customers: Smart grids deliver a plethora of advantages to customers through the implementation of interactive and scalable models of power grid and energy demand. Customers, encompassing both residential and industrial users of electricity, experience a range of benefits:

- Empowered Energy Options: Customers, now often prosumers producing their own electricity through alternative methods like solar energy, biomass, and wind, benefit from the flexibility offered by smart grids.
- Real-time Interaction: Real-time communication with smart grid control and monitoring systems facilitates precise measurement and optimization of customers' energy utilization within the grid.
- Enhanced Control: The deployment of smart meters and other smart grid equipment empowers consumers to actively control their energy consumption in real time. This capability enables them to avoid peak loads, leveraging associated price benefits.
- Cost Savings: Customers can strategically schedule high-energy-consuming activities, such as operating washing machines, dryers, and dishwashers, during off-peak times when energy prices are significantly lower. This not only results in cost savings for customers but also contributes to a more efficient use of energy resources.
- Reduced Generation Capacity Demand: By optimizing their consumption patterns, customers not only save money but also contribute to a reduced demand for generation capacity during peak periods, fostering a more sustainable and balanced energy grid.

2.2 Smart grid systems

Smart grids rely on advanced and contemporary communication and information infrastructure to enhance energy production, distribution, and storage, consequently streamlining the cost and effort involved in management and planning. Our focus will predominantly be on the information systems integral to the management of data. This emphasis underscores the pivotal role of robust and efficient information systems in ensuring the seamless operation and optimization of smart grid functionalities.

Communication System: plays a critical role, and it should maintain high bandwidth capacity and speed, data security and privacy, etc. Data transmission in smart grid systems is based on communication technologies, starting with access network technologies including PLC, ZigBee, WIFI, etc., followed by area network technologies, using M2M, cellular networks, Ethernet, etc. Subsequently, core network technologies including IP, IMPLS etc., are utilized. Finally, backbone network technologies rely on fiber technologies, microwave link, IP-based wavelength, network, and other optical technologies, for more details about data transmission see subsection [11].

Information Systems: Information systems constitute vital elements within smart grids, working collaboratively to establish a flexible, scalable, and efficient grid, as depicted in **Figure 4**. These utility information systems play a crucial role in controlling and processing data sourced from utility substations and various electricity consumers, including those in commercial, residential, and industrial sectors. The information extracted is then utilized to derive valuable insights into the state of grid lines and equipment, energy consumption, consumption patterns, and more. Utility information systems encompass several key components:

- Supervisory Control and Data Acquisition (SCADA) System [12]: This system collects data from the utility field and manages the electrical grid infrastructure. It communicates with other information systems to provide comprehensive reports about the network.

- Customer Information System (CIS): CIS processes data from electricity consumers, fostering effective communication and data exchange among various information systems.

- Geographic Information System (GIS): GIS aids in processing data related to the geographical aspects of the grid, enhancing spatial understanding and coordination.

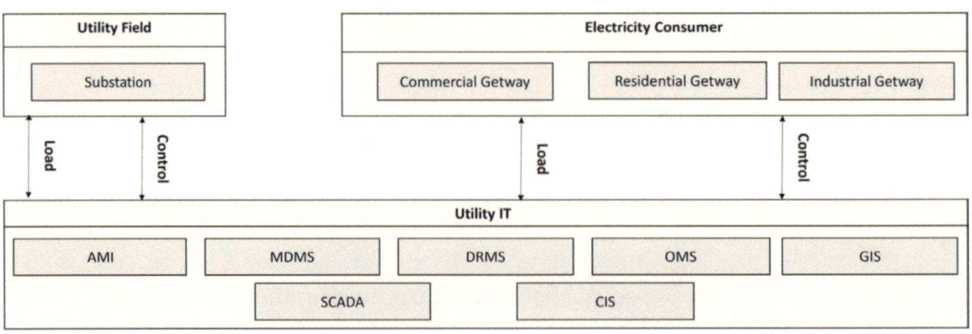

Figure 4.
Information systems in smart grid.

- Advanced Metering Infrastructure (AMI) [13]: AMI plays a crucial role in collecting and managing data from smart meters, facilitating real-time monitoring of energy consumption.

- Meter Data Management System (MDMS): MDMS processes and manages data collected from meters, ensuring accurate and efficient handling of consumption information.

- Demand Response Management System (DRMS): DRMS interacts with all other systems, ensuring effective coordination and response to demand fluctuations, ultimately contributing to grid stability.

- Outage Management System (OMS): OMS is instrumental in managing and responding to outages promptly. Together with DRMS, it ensures a comprehensive view of the grid and enhances overall consumer satisfaction.

These systems collectively provide a holistic perspective on the grid and consumer interactions, emphasizing the interconnected nature of information systems within the smart grid infrastructure. Further details about the functionalities and roles of each system will be elaborated below.

2.3 Data management issues in smart grid

Smart grid systems generate substantial amounts of data, with systems like SCADA collecting data every 2–5 seconds and AMI systems capturing data at intervals of 1–15 minutes. Consequently, utilities are confronted with a multitude of challenges spanning from strategic considerations to operational performance in the realm of data management. The sheer volume and frequency of data acquisition pose complex challenges that utilities must navigate to effectively harness and leverage the wealth of information generated by smart grid systems. These challenges encompass strategic decision-making regarding data utilization, the establishment of robust data management strategies, and the optimization of performance in handling and analyzing the copious amounts of data flowing through the smart grid infrastructure. As utilities grapple with these challenges, finding innovative and efficient solutions becomes imperative to ensure the continued success and advancement of smart grid technologies.

Standards and Interoperability: The smart grid represents a heterogeneous and intricate environment characterized by diverse devices, networks, systems, and data types. Within this framework, variations abound, encompassing networks with differing processing speeds, devices with or without energy constraints, interactive or non-interactive systems, and continuous or non-continuous data, among others. This diversity imposes a range of requirements and challenges on smart grids concerning data integration, including bandwidth constraints, error management, limited resources, and the demand for high scalability. In the absence of standardized protocols, utilities grapple with the use of disparate protocols featuring distinct definitions and communication techniques, making interoperability an arduous task. To address this challenge and establish standardization in smart grids, various information models have been developed. Notably, IEC 61850 serves as a communication standard MDMS and related enterprise applications. The IEC 61970/61968 Common Information Models (CIM) relies on IEC 61850 as the foundation for information exchanges and messaging. The evolution of smart grid technology has led to the integration of advanced protocols, such as IEC 61850–90-7, specifically tailored for smart inverters. Additionally, protocols like IEEE 1815 (DNP3) and IEEE 2030.5 (SEP2) offer advanced communication capabilities, allowing the utilization of existing communication infrastructures. These standardized protocols play a crucial role in facilitating seamless communication, interoperability, and efficient data management within the complex and dynamic landscape of smart grids [10].

Management of Massive Data Volume: The implementation of smart grids highlights the significant costs associated with storing and processing the substantial volume of data essential for effective grid management. Regrettably, many utilities face challenges in fully leveraging the wealth of new data collected due to limitations in infrastructure and/or insufficient data analysis skills. This underutilization of data not only hampers the potential benefits for utilities but also poses challenges for customer data management. Beyond the considerable data management issues utilities encounter, there are additional challenges pertaining to customer data. The untapped potential of this data diminishes the opportunities for consumers to actively participate in the smart grid, specifically in terms of controlling their energy consumption and capitalizing on price benefits by avoiding peak loads. To harness the full advantages of smart grids, consumers need to be cognizant of the extensive data that can be derived through these platforms. Customers should be educated about how to effectively engage with their meters and various platforms within the grid. This knowledge empowers them to make informed choices that contribute to energy conservation and result in cost savings. By raising awareness about the potential of smart grid data and providing consumers with the necessary skills, a more engaged and empowered consumer base can emerge, unlocking the true potential of smart grid technologies.

Security and Data Privacy: Within the intricate ecosystem of the smart grid, millions of interconnected devices communicate through networks, exposing the grid to potential vulnerabilities. The adoption of virtualization technologies, a cornerstone for leveraging cloud computing, allows electrical companies to operate applications in virtual machines, leading to reduced investment costs in terms of hardware and energy [14]. However, this approach introduces security challenges due to the shared platform among multiple users. Furthermore, network bandwidth emerges as a significant challenge, leading to low-latency issues in real-time applications that necessitate highly scalable, available, and fault-tolerant connections [15]. These information and ICT dimensions escalate the risks associated with compromising the security objectives of the smart grid. These objectives include ensuring availability, integrity, confidentiality, and accountability. The confidentiality of data becomes a concern when there is no secure connectivity between devices [16]. Establishing secure communication channels requires the adoption of authentication mechanisms [17]. The "authentication, authorization, and accounting" (AAA) mechanism is commonly employed for this purpose. Authentication verifies users through credentials, authorization delineates individual user permissions, and accounting oversees user activities [18]. These security measures are imperative for safeguarding the integrity and reliability of the smart grid communication infrastructure.

Data Availability: is paramount in smart grid communications, ensuring uninterrupted electricity delivery and system functionality. Defending against attacks like false data injections is crucial, requiring real-time detection and mitigation mechanisms. The most used mechanism for ensuring data availability in smart grid systems is redundancy and fault tolerance [19, 20]. Redundancy involves duplicating critical components or data across multiple systems to mitigate the impact of failures or disruptions. Fault-tolerant techniques are also commonly employed to ensure continuous operation in the event of component failures or network disruptions.

Data Integrity: is paramount to the security of the smart grid infrastructure. Data integrity measures ensure that information remains unchanged and uncorrupted during transmission or storage. Robust encryption algorithms and digital signatures are commonly used to maintain data integrity, providing assurance against unauthorized alterations, or tampering [21, 22].

Data Accountability: in smart grid systems refers to the principle of holding individuals or entities responsible for their actions and decisions within the grid infrastructure. It encompasses various aspects, including ensuring transparency, traceability, and responsibility for system operations, data management, and security measures. The above-discussed authentication mechanism is the most used to verify the identities of users and devices accessing the system.

3. Big data for smart grid

3.1 Big data life cycle for smart grid systems

The term "big data" encompasses more than just the sheer quantity of datasets. Beyond (1) volume, it is characterized by (2) variety, representing diverse data formats (structured, semi-structured, or unstructured), (3) velocity, ensuring timely processing, (4) value, allowing the extraction of meaningful insights from collected datasets, (5) variability, accounting for the inconsistency in data, and (6) veracity, emphasizing the trustworthiness of the data. **Figure 5** illustrates big data technologies for smart grids.

- Data Source: In the realm of smart grids, a diverse array of data sources contributes to a comprehensive understanding of the electrical infrastructure. These data sources can be categorized into distinct classes based on the nature of the information they provide. Firstly, operational data encompasses critical electrical details, including real and reactive power flows, demand response capacity, and voltage levels. Secondly, non-operational data is unrelated to the power grid's immediate functioning but includes essential information like master data, power quality, and reliability metrics. Another crucial category is

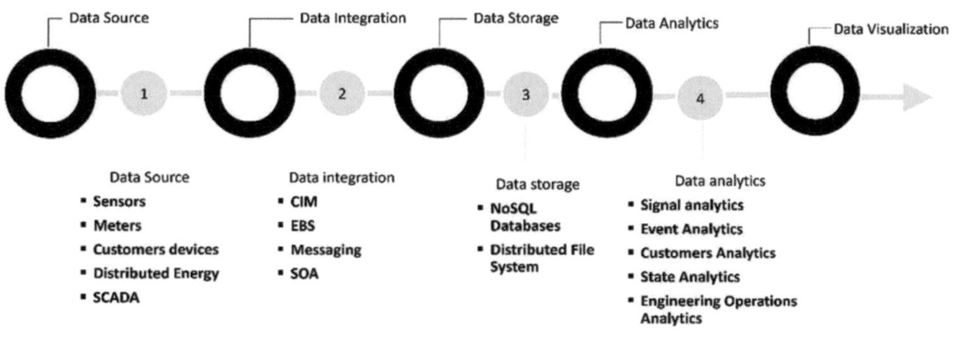

Figure 5.
Big data architecture for smart grid.

meter usage data, offering insights into power consumption patterns, such as average usage, peak demand, and time-of-day variations. Event message data, the fourth type, originates from events detected by smart grid devices, such as voltage fluctuations or fault detection events. Lastly, metadata serves as a unifying element, facilitating the organization and interpretation of all other data types. These diverse data streams are sourced from a variety of points, including meters, sensors, devices, substations, mobile data terminals, control devices, intelligent electronic devices, distributed energy resources, customer devices, and historical data.

- Data Integration: In the current landscape, the enhancement of smart grid reliability, persistence, efficiency, and performance is driven by the utilization of modern information and communication technologies along with advanced operational approaches. To ensure seamless data integration, various technologies are employed [23]:

  - Service Oriented Architecture (SOA): The multitude of software in enterprise systems often poses a challenge in terms of management. SOA addresses this issue by establishing a unified approach for software communication, making data integration more adaptable and streamlined. In the context of smart grids, SOA is particularly employed in on-demand systems.

  - Enterprise Service Bus (ESB): ESB facilitates communication management between diverse systems like GIS, OMS, CIS, etc. This technology yields numerous benefits, including cost reduction, time efficiency, and improved integration management and monitoring. In smart grids, ESB technologies are closely intertwined with SOA, reinforcing robustness, and flexibility.

  - Common Information Models (CIM): CIM serves a critical role in smart grid persistence and integrated data architecture. Specifically designed for the electric power industry using UML models, CIM significantly contributes to energy management systems by ensuring data integration efficiency in terms of time and cost. In power systems, CIM is indispensable for ensuring data interoperability, especially when implementing various applications. Operating at the data transformation level, CIM collaborates with ESB to normalize and standardize data between smart grid systems.

  - Messaging: Messaging systems operate on the exchange of messages containing data and other information among various applications, managed by a messaging server. This approach enhances communication efficiency and is integral to the overall data management strategy.

In essence, these technologies and approaches play a pivotal role in optimizing smart grid operations, ensuring seamless communication, and promoting data interoperability across diverse systems.

- Data Storage: Data storage plays a pivotal role in the smart grid, acting as a crucial link between the collection of data from various sources and the delivery of that data to analytics tools, necessitating fast input/output operations per second (IOPS).

The requirements of handling big data necessitate a sophisticated and scalable data storage mechanism. One prevalent solution is the use of distributed file systems (DFS), which facilitates file sharing and storage resource access across multiple users and machines. Operating on a client/server storage mechanism, DFS allows each user to obtain a local copy of the stored data. Numerous solutions, including Google's GFS, Quantcast File System, HDFS, Ceph, Luster, GlusterFS, and PVFS, leverage DFS for effective data storage. Another innovative approach to data storage is the utilization of NoSQL databases, offering a fresh perspective to address the limitations of traditional relational SQL databases when dealing with massive datasets. NoSQL databases are characterized by three primary architectures: key-value solutions like Dynamo and Voldemort, column-oriented solutions including Cassandra and HBase, and document database solutions such as MongoDB and CouchDB. These diverse architectures provide flexibility and efficiency in managing and storing large volumes of data in the context of smart grid applications.

- Data analytics: In the smart grid ecosystem, data collection from diverse sources is extensive, resulting in a substantial dataset that is essential for analytics consumption. Analytics plays a pivotal role in augmenting the intelligence, efficiency, and profitability of the grid, with various types of analytics including signal analytics, event analytics, state analytics, engineering operations analytics, and customer analytics. Currently, several models integrate these diverse analytics classes, including descriptive, diagnostic, predictive, and prescriptive models. Descriptive models elucidate customer behaviors in demand response programs, offering a fundamental understanding of their practices. Following customer description, diagnostic models delve deeper to comprehend specific behaviors and analyze individual decisions. These models collectively contribute to predictive models that forecast future customer decisions. At the apex of smart grid analytics are prescriptive models, influencing marketing strategies, engagement tactics, and decision-making processes [24]. The processing of big data in smart grids can occur through two primary approaches. Batch processing handles data within specific time intervals and is suited for scenarios with relaxed response time requirements. On the other hand, stream processing is employed for real-time applications, demanding minimal latency in response time for effective and timely decision-making.

- Data Visualization: Effective data visualization is integral to the assessment of smart grids. Various techniques, particularly multivariate high-dimensional visualization, play a significant role in presenting complex data. These techniques, including 2D and 3D visualization, address the challenge posed by the multitude of variables in smart grids. Visualization tools such as Scatter diagrams, parallel coordinates, and Andrew curves are employed to simplify the representation of high-dimensional data, offering clarity in the assessment of smart grid information, including 3D Power-maps [25].

- Data Transmission: Data transmission is a critical aspect of big data, influencing every phase of data processing. It is essential to maintain high bandwidth capacity and speed and ensure data security and privacy. In the context of smart grids, data transmission relies on various communication technologies outlined in "communication systems". This includes access network technologies such as

PLC, ZigBee, and WIFI, followed by area network technologies involving M2M, cellular networks, and Ethernet. Core network technologies utilize IP, IMPLS, while backbone network technologies leverage fiber technologies, microwave links, IP-based wavelength division multiplexing (WDM) networks, and other optical technologies. The robustness and efficiency of data transmission systems are crucial for the seamless operation of smart grids.

3.2 Criteria for choosing big data technologies

Big data technologies propose a variety of tools, prompting utilities to carefully select platforms and tools aligned with their specific goals. As highlighted in the preceding subsections, the big data life cycle encompasses five fundamental phases: data sources, data integration, data storage, data analytics, and data visualization. Notably, big data analytics stands out as the most crucial step in this life cycle. Consequently, by understanding the analytical processes involved, utilities can strategically determine the types of data to acquire, ascertain how to store it efficiently, and even select appropriate visualization techniques to enhance their insights. This strategic approach ensures that utilities derive meaningful knowledge from the vast datasets encountered in the big data landscape.

When electrical companies embark on selecting solutions, a multitude of factors must be carefully considered. Criteria such as computation speed, compatibility, graphic capabilities, cloud compatibility, and more come into play. To navigate through these considerations, utilities often leverage multiple criteria decision making (MCDM) tools. Among these tools, the analytic hierarchy process (AHP) stands out as one of the most popular MCDM methods, valued for its ability to incorporate both quantitative and qualitative considerations.

In decision-making applications, AHP proves beneficial for selecting big data analytics platforms. This involves defining criteria across technical, social, cost, and policy perspectives. The AHP model facilitates a comprehensive evaluation of technical perspectives, encompassing hardware and resource configuration requirements. **Table 1**, as described in [26], outlines the technical perspectives relevant to big data analytics platform selection, providing a structured framework for decision-making in this complex landscape.

Technical perspective	Criteria
Availability and fault tolerance	Redundancy and resilience in networks, servers, physical storage, etc.
Scalability and flexibility	Tools must be evolutionary and scalable
Performance (latency)	Data processing time (single transaction and query request)
Computational complexity	Computation tools extension (data mining and business intelligence)
Distributed storage capacity and configurations	Storage systems parameters such as storage nodes needed in terms of availability, periodic basis, etc.
Data processing modes	Batch, real, and hybrid processing
Data security	Security compliance according to the platform requirements

Table 1.
AHP model technical perspective.

3.3 Big data implementation in smart grid

The emphasis will be placed on customer data analytics, given its association with the concept of smart consumers. In this paradigm, consumers are not only end-users but potential producers of clean energy, active participants in their consumption patterns, and pivotal actors in achieving a balance between production and consumption. The realm of customer data analytics presents a significant opportunity for utilities to gain profound insights into customer behavior. This understanding serves as a foundation for making strategic decisions that align with the evolving dynamics of smart grids and the active involvement of consumers in the energy ecosystem.

- **Added Value of Customer Data analytics:** The implementation of big data analytics for customer data has shifted from being an option to a necessity for electrical companies. Consumers are actively participating in smart grids through the utilization of smart meters, granting them enhanced control over their energy consumption. Utilities leverage demand response (DR) programs to access real-time data on demand curves across various consumption points, enabling more precise calibration and forecasting. This, in turn, facilitates the more efficient regulation of production curves according to demand, minimizing losses from overproduction. The benefits of DR extend to real-time diagnostics of meters and equipment near consumers, triggering alarms, and executing self-healing systems. One of the motivations behind DR is the improvement of customer engagement, enabling utilities to interact with customers' energy needs even during power outages. Dynamic pricing, integrated into DR, encourages consumers to monitor and adapt their usage in real time, particularly during peak periods, offering them the flexibility to align their consumption with cost-saving measures [13]. The successful implementation of these initiatives relies heavily on employing customer data analytics techniques, as illustrated in **Table 2**.

- **Big Data Tools for Customer Data Analytics:** Customer data, often measured in terabytes and presented in diverse formats, demands high velocity, scalability, and fault tolerance in its processing, storage, and visualization. Implementing big data solutions involves leveraging various tools, with analytics tools standing out as crucial elements in business decision-making. The diverse sources of customer data, ranging from smart meters to devices and historical data, necessitate the use

Analytics type	Methods	Example of use
Descriptive	• Data mining • Data aggregation	Demand response Programs
Diagnostic	• Correlation • Cause-and-effect analysis	Analyze customer behavior and decisions
Predictive	• Statistical models • Forecasts techniques	Predict customer decision
Prescriptive	• Business rules • Machine learning • Computational modeling procedures	Marketing models and engagement strategies

Table 2.
Customer data analytics.

of integration tools to ensure data uniformity. Messaging tools prove to be highly efficient for integrating raw data and are thus aptly employed for customer data integration. When it comes to big data analytics, several processing modes can be employed:

- Batch Processing Tools: Big data analytics offers various methods for data processing, with batch processing being a significant approach. Hadoop emerges as a suitable choice for batch analytics in the context of smart grids. Given the geographical distribution of smart grid systems, the presence of distributed file systems becomes invaluable. Hadoop encompasses Hbase as a database system, Hadoop distributed file system (HDFS) as a storage system, and MapReduce as a processing engine. However, Hadoop has limitations in handling modern information technology (IT) systems with respect to data velocity, scalability, and machine learning algorithms [27].

- Real-Time Processing Tools: Real-time processing tools excel in faster execution, particularly when dealing with data possessing higher velocity requirements. These tools employ stream processing or complex event processing systems. Notable solutions for real-time processing in smart grids include S4, Splunk, and Storm. Storm stands out as a fitting choice due to its attributes as an open-source, distributed, and fault-tolerant system. Storm offers numerous advantages for real-time processing in smart grids, including reliable message handling, parallel computations, and a simple programming model. Storm can be seamlessly integrated with Kafka for data integration and Hbase for data storage.

- Hybrid Processing Tools: Hybrid processing tools adeptly handle both batch and real-time processing. Spark, a versatile framework, is employed for batch processing and provides a real-time processing solution through Spark Streaming. Spark is equipped for large-scale data processing and includes useful tools like Spark SQL, Spark Streaming, a machine learning library, and GraphX. These features make Spark well-suited for meeting big data requirements in smart grids. Spark Streaming utilizes a real-time complex event processing engine to address velocity issues. For data storage in a Spark environment, HDFS or even Hbase can be utilized [28]. Another framework, Apache Flink, excels in processing data in both batch and stream modes. Flink boasts extensive APIs for transformation functions (map/reduce, group, etc.), making it scalable, easy to deploy, fault-tolerant, and fast in execution. Flink's efficiency in machine learning is enhanced by its dedicated machine learning library, FlinkML. With built-in libraries for accessing HDFS, Flink integrates seamlessly with HDFS for data storage.

4. Forecasting energy consumption in buildings

4.1 Forecasting approaches

In fact, forecasting is used in several domains for many purposes, and each use has its own horizon and granularity depending on the nature of data used and the desired results. Forecasting uses high degree of probability and statistical techniques to predict future and unknown events, so it relies on modeling to analyze current and historical facts [29]. In general, energy forecasting models can be classified into three categories [30–32]:

- White box models are based on detailed physical information and use known and conventional knowledge like physics equations to describe cases. It is not easy to apply it, and it takes more time to run despite those limits, this kind of forecasting gives more accuracy. This type of models is also known as physics-based models, forward classical approach, calibrated simulation approach or as engineering methods.

- Black box models are based on historical data and uses data mining techniques, statistical analyses, and machine learning algorithms to get the relation between the input and the future outputs values. It Is not easy to apply it, but it presents a high running speed with good accuracy. This type of models is also known as intelligent techniques, statistical analyses, data-driven techniques, or as time series techniques.

- Gray-box models are based on the combination of both white-box and black-box models, by improving single data-driven techniques with optimization methods or combining several machine learning algorithms. It is not easy to apply it, and it takes more time to run but it presents a high accuracy. This type of models is also known as hybrid models, gray-box models, or improved models.

For this chapter, we will focus on black-box models, to describe this forecasting process, its concept as well as the benchmark of the models that offer.

4.2 Data-driven approach

Data-driven methods always seek to identify the relationship between inputs and outputs by applying data mining techniques, statistical analyses, and machine learning algorithms. This process goes through three stages, training to run training datasets and produce results, validation to run evaluation datasets that are different from the training datasets which aim assessing the performance of the implemented algorithm and finally the testing step to evaluate the forecasting model performances using accuracy metrics. This type of techniques can be divided into two categories [30, 33]:

- **One Dimensional Data-driven Model:** One dimensional data-driven model is a simple and basic technique for forecasting problems, because it implements only one predictive algorithm. The most used algorithms for building forecasting are as follows:

    - Linear Regression (LR): is an algorithm to estimate the relationship between a dependent and an independent variable using statistical manner, which makes it suitable for prediction and forecasting applications [34]. This model sets the linear relationship between independent variables, several inputs (x) and single output (y). This algorithm is very useful to figure out the cause-and-effect relationship between variables, and the main goal of the linear regression is to find the optimal values for *a* and *b* in the linear equation **y = a + b*x** [34].

- Isotonic Regression (IR): belongs also to regression algorithms, but compared to linear regression this algorithm does not follow any form for the target function. Isotonic regression minimizes the mean squared error on the training data to have non-decreasing approximation of a function. The main purpose of this algorithm is to find real values based on observed response and X = x1, x2, ... ,xn of the training data. Formally isotonic regression uses a unique function called isotonic regression [35]. In general, the unknown values are calculated using this function:

$$f(x) = \sum_{k=1}^{n} \text{wk}\left(\text{yk} - \text{xk}\right)^2 \tag{1}$$

- Decision Tree (DT): is an approach based on tree architecture that lists all possibilities under defined constraints, and it belongs to supervised learning. Decision tree algorithms are used to predict using continuous data or to classify using discrete data [36].

- Support Vector Regression (SVR): uses the same principle as support vector machine, but for regression problems, SVR is accomplished by introducing an e-insensitive region around the function, called the e-tube. This tube reformulates the optimization problem to find the tube that best approximates the continuous-valued function, while balancing model complexity and prediction error. More specifically, SVR is formulated as an optimization problem by first defining a convex e-insensitive loss function to be minimized and thus finding the flattest tube that contains most of the training instances [37].

- Gradient Boosting Trees (GBT): is based on three concepts, function loss to be optimized which can be squared error for regression problems or logarithmic loss for classification problems. Weak learner to make predictions and additive model to add weak learners to minimize the function loss [38].

- Artificial Neural Networks (ANN): uses multiple layers of nodes which are related to each other, and these layers are divided into input layers, hidden layers, and output layers. At artificial neural networks, there is exactly one input layer with a size equal to the length of the feature vector and one output layer equal to the number of labels. For hidden layers, one is sufficient for a great number of problems be more than one can increase the training time [39].

- **Multi-Dimensional Data-driven Model:** Multi-dimensional data-driven model is a complicated and advanced technique for forecasting problems because it can combine multiple models to implement a hybrid forecasting.

  - Ensemble models: mix several single algorithms together, and to do this combination ensemble models apply two strategies: (i) homogeneous modeling and (ii) heterogeneous modeling. Random forest is among this king of black-box models, and it is a regression algorithm based on ruining multiple decision trees in parallel with training time to overcome the overfitting.

- Improved models: are the result of combining single models and optimization techniques to improve the prediction accuracy. Swarm intelligence algorithms such as particle swarm optimization and genetic algorithms are among the optimization techniques used.

4.3 Machine learning for data-driven models

The aim of machine learning concept is to automate and mechanize the acquisition of knowledge from experience. Furthermore, machine learning improves performance of computational methods and algorithms. Thus, machine learning is hardly linked to the data-driven model which finds relationships among the state variables of the system without using physical behavior of the system. Machine learning uses some approaches to select and extract features from input datasets to train data-driven models including supervised, unsupervised, reinforcement, and transfer learning. **Table 3** describes some machine learning approaches that are associated with data-driven models. In general, to apply both machine learning and data-driven models to a problem, certain steps must be followed, starting with defining objectives then collecting, exploring, processing, and visualizing data and finally running, evaluating, and comparing performance of the model.

5. Review of forecasting electrical consumption in educational buildings

5.1 Data selection and preparation

To improve prediction accuracy, the system must collect all relevant data with high impact on school building energy consumption. Electrical consumption monitoring in a building can be less accurate, if it does not include all factors that hardly affect the energy use. Thus, focusing on the most relevant information and eliminating the irrelevant ones is a central step in forecasting problems. For concreteness,

Task	Type	Data-driven model
Classification	Supervised	• Artificial neural networks • Support vector machines • Random forest
Regression	Supervised	• Linear regression • Decision tree • Support vector regression
Clustering	Unsupervised	• Gaussian mixture models • Density • Based on spatial clustering • Random forest
Anomaly Detection	Unsupervised	• Support vector machines

Table 3.
Machine learning uses with data-driven models [40, 41].

	[42]	[43]	[44]	[45]	[46]	[47]	[48]
Temperature	X		X	X	X	X	X
Irradiation	X					X	
Humidity	X	X				X	
Wind speed	X				X	X	
Electricity usage	X	X	X	X	X	X	X
Building type	X						
Day of the week	X	X			X		
Day of the year	X	X					
Hour of the day		X					
Event day		X					
Event type		X					

Table 4.
Identification of relevant data of reviewed research papers.

we reviewed some studies about forecasting electrical consumption for education institutions to highlight the more relevant data to be collect in this context. **Table 4** summarizes for each study the data used to forecast electrical consumption. In general, most of these studies use temperature, humidity, and wind speed to represent meteorological data. But a few works use the occupancy data for building electrical forecasting, and even those studies which use it do not give enough details.

As a result, many studies have highlighted the significance of occupancy and weather data in electricity consumption prediction. However, most of these works either do not integrate these two main data types or do not use detailed occupancy data. This work explores the use of both occupancy and meteorological data, and the study gives a daily occupancy more detailed than other works, which will ensure the accuracy and the quality of predictions. The proposed solution uses in addition to electrical consumption:

- **Occupancy data:** presents time data, which is strongly linked to daily occupancy and schedule data based on holidays, school year, semester period, and even weekends. In fact, the occupancy behaviors and activities affect the energy consumption patterns.

- **Meteorological data:** presents many factors, which must be taken into consideration because it strongly affects occupancy behavior and materials consumption. Temperature, humidity, wind speed, and pressure are all main factors that should be selected as input data to implement our forecasting model. These data can be collected from several devices such as sensors, weather data, and meteorological station data.

5.2 Model selection and training

According to Runge et al. study [49], in energy consumption forecasting 84% of research use artificial neural network (ANN) models applied with black-box-based models, followed by white-box with 12% and finally gray-box models with 4%. Thus, for this work, we will focus on black-box models, to describe this forecasting process, its concept as well as the benchmark of the models that is offered.

The work of Bourdeau et al. [30] presents a review on data-driven building energy modeling techniques. They introduce the most prevalent techniques and provide an up-to-date overview of recent studies and advancements in building energy consumption modeling and forecast studies. According to Bourdeau et al. works, the number of reviewed research papers from 2007 to 2019 for supervised machine learning for building energy consumption modeling and forecast show that ANN is the most studied, then in the second place came the support vector machine (SVM), and in the third place came the regression models. Wei et al. [50] also review the prevailing data-driven approaches used in building energy analysis. Wei et al. review many methods for prediction building energy including artificial neural networks, support vector machines, statistical regression, DT, and genetic algorithm. They conclude that ANN, SVM, and regression techniques are the most used in these cases. In the same context, Amasyali et al. [51] make an overview of the most used algorithms for energy consumption in buildings. They conclude that an overall 47% of the energy consumption prediction models utilized ANN as machine learning algorithms, while 25% used SVM, 4% DT, and 24% other statistical models.

Some studies implement their own systems and run many algorithms on them to find the most appropriate model in the case of electrical consumption forecasting in the buildings. Grolinger et al. [43] explore many prediction intervals for electrical consumption in the context of event-organizing venues including daily, hourly, and 15-min. Grolinger et al. compare forecasting results accuracy for two machine-learning approaches, ANN and SVM. They achieved high consumption prediction accuracy with daily data better than hourly or 15-min readings and using the ANN model instead of the SVM model. The work of Amberet al. [52] is also interested in daily forecasting electricity consumption of building. Amberet al. compare prediction capabilities of five different intelligent system techniques including multiple regression (MR), genetic programming (GP), ANN, deep neural network (DNN), and SVM. Similarly, to Grolinger et al., Amberet al. results demonstrate that ANN performs better than all the other four techniques. Kim et al. [53] study compares building electric energy prediction approaches that use a traditional statistical method (linear regression) and ANN algorithms. Kim et al. results illustrate that the ANN modeling was more accurate and stable than the linear regression method. As a result, the majority of studies find that ANN are the primary models employed to evaluate and predict energy consumption [43, 52, 54, 55].

ANN offers great number of models, but according to Runge et al. [49] study most research uses multi-layer perceptron (MLP). Runge et al. review shows that 61% of the ANN models use MLP with a large portion using a two hidden layers. The result of the analysis showed that MLP is followed by radial basis neural networks, non-linear autoregression neural network, general regression neural network, and nonlinear autoregressive network with exogenous inputs neural network with 2–7% each. MLP is regression artificial neural network that has been widely used in the building sector for supervised learning using ANN. Chammas et al. [56] propose a system based on MLP to predict energy consumption of a building. They compare their system against four other algorithms, namely linear regression (LR), SVM, gradient boosting machine (GBM), and random forest (RF). Chammas et al. achieve that MLP is the best system configuration energy consumption forecasting in a building. Wahid et al. [57] find also that MLP is the best model to predict electrical consumption in building.

Wahid et al. study compares MLP and RF, and their results show that MLP achieved 95.00% accurate result, whereas the accuracy observed by RF was only 90.83%.

All these works listed above like this article compare the efficiency of ANN to other machine learning models in the case of building power consumption forecasting. However, our work compares MLP to many models that are the most used in this field. Thus, this research will give more precision on the performance of the MLP.

5.3 Model testing and validation

Accuracy metrics help to validate forecasting performances of data-driven algorithms. In fact, validation step is used to verify the quality of the model, and error metrics are used to measure the difference between values predicted and the values observed. Many metrics exist, but the choice of the good one depends on the use and the objective of the work:

- **Mean Absolute Error (MAE)** [58]: this model evaluation metric describes the average of the absolute values of all differences between the forecast and the real values expressing the same phenomenon.

$$MAE = \frac{\sum_{0}^{N} |Y\ real - Y\ pred|}{N} \tag{2}$$

where N is the total number of data points.

- **Normalized Mean Absolute Error (NMAE)** [59]: this metric presents the MAE by the mean of the time series; we can interpret the result as a weighted mean absolute percentage error.

$$NMAE = \frac{MAE}{mean(Y\ real)} * 100 \tag{3}$$

- **Mean Square Error (MSE)** [60]: this model evaluation metric presents the mean of the squared prediction errors over all instances in the test set. In fact, it describes the difference between the real and the predicted results for an instance. Mean squared error presents how the predicted results are close to real set of points. In general, the smaller value of this metric is the most accurate.

$$MSE = \frac{\sum_{1}^{N} (Y\ real - Y\ pred)2}{N} \tag{4}$$

- **Root Mean Square Error (RMSE)** [61]: presents the concentration of data around the line of the best fit. Root mean squared error is usually used in regression analysis for numerical predictions because it is a good measure of accuracy in the case of comparing prediction errors of different models or model configurations for a particular variable.

$$MSE = \sqrt{MSE} \tag{5}$$

- **R-squared (R2)** [62]: also called coefficient of determination, is a statistical measure that shows how close are the data to the fitted regression line. It indicates the percentage of the variance in the dependent variable that the independent variables explain collectively. R-squared measure is a percentage value between 0 and 100% scale. In general, the higher the R-squared, the better the model fits the data.

$$R2 = 1 - \frac{\sum_1^N (Yreal - Ypred)2}{\sum_1^N (Yreal - Ymean)2} \tag{6}$$

Bourdeau et al. [30] propose an overview study of the most used metrics based on reviewing many studies. They find that the RMSE, the coefficient of variation of RMSE (CV-RMSE), and the MAE assessed in 47, 38 and 36%, respectively. On the other hand, the R2, the MSE, the mean relative error (MRE), the mean bias error (MBE), and the normalized mean bias error (NMBE) are used only for 27, 16, 9, 2, and 4%, respectively. Zhang et al. [63] study shows that CV-RMSE must be the first performance measure to be selected followed by other metrics. Zhang et al. consider CV-RMSE is more important metric followed by RMSE, but if these two values were unavailable, then MAPE was selected. If MAPE was unavailable, then R2 was selected. If R2 was unavailable, then the most relevant error method presented was selected and indicated as others. However, this order is not usually respected, Runge et al. [49] have done a review of the most used error metrics, and they find that MAPE is predominately (38%) used as the main performance measure within forecasting papers, with CV-RMSE and R2 accounting for 17–20% of the performance metrics applied.

6. Conclusions

In conclusion, the integration of big data and analytics in the context of smart grids, with a specific focus on predicting energy consumption in educational institutions, represents a transformative leap toward sustainable and efficient energy management. The multifaceted application of predictive analytics, leveraging advanced machine learning models has demonstrated promising results in forecasting energy consumption patterns.

The comprehensive evaluation of these models across various spatial uses within educational institutions has shed light on their strengths and limitations. The insights gained from this study provide a foundation for guiding future applications in energy consumption prediction. As educational institutions increasingly adopt smart grid technologies, and the lessons learned from this research contribute to the development of an effective strategies for optimizing energy usage. Furthermore, the exploration of data sources, storage mechanisms, and transmission technologies within the smart grid ecosystem underscores the complexity of the predictive analytics life cycle. The role of information systems, from SCADA to demand response management, is pivotal in ensuring seamless data integration and efficient energy management.

As the demand for energy-efficient operations continues to grow, the necessity for accurate and adaptable predictive models becomes paramount. This chapter not only highlights the significance of integrating big data analytics but also emphasizes the need for careful consideration in choosing the right tools and platforms. The utilization of multiple criteria decision making (MCDM) tools, such as the analytic hierarchy process (AHP), emerges as a valuable approach for making informed decisions based on technical, social, cost, and policy perspectives. In essence, the predictive analytics landscape in smart grids is dynamic and evolving, offering a plethora of opportunities for enhancing energy efficiency. The findings and considerations presented in this chapter contribute to a broader understanding of the predictive analytics ecosystem in smart grids, laying the groundwork for continued advancements in energy consumption prediction, particularly within the educational sector.

Conflict of interest

The authors declare no conflict of interest.

Appendices and nomenclature

AAA	authentication, authorization, and accounting
AHP	analytic hierarchy process
AMI	advanced metering infrastructure
ANN	artificial neural networks
CIM	common information models
CIS	customer information system
DFS	distributed file systems
DR	demand response
DRMS	demand response management system
DT	decision tree
GBT	gradient boosting trees
GIS	geographic information system
HDFS	Hadoop distributed file system
ICT	information and communication technologies
IOPS	input/output operations per second
IR	isotonic regression
IT	information technology
LR	linear regression
MAE	mean average error
MCDM	multiple criteria decision making
MDMS	meter data management system
MSE	mean square error
NMAE	normalized mean absolute error
NMBE	normalized mean bias error
OMS	outage management system
R2	coefficient of determination
RF	random forest
RMSE	root mean square error
SCADA	supervisory control and data acquisition
SVM	support vector machine
SVR	support vector regression
WDM	wavelength division multiplexing

Author details

Houda Daki, Basma Saad, Asmaa El Hannani, Abdelfatteh Haidine*
and Hassan Ouahmane
Laboratory of Information Technologies, National School of Applied Sciences,
Chouaib Doukkali University, El Jadida, Morocco

*Address all correspondence to: haidine.a@ucd.ac.ma

References

[1] Agarwal V, Lefteri HT. Smart grids: importance of power quality. In: Energy-Efficient Computing and Networking: First International Conference. Athens, Greece; 2011

[2] Amin SM. Smart grid: Overview, issues and opportunities. advances and challenges in sensing, modeling, simulation, optimization and control. European Journal of Control. 2011; **17**(5-6):547-567

[3] Yan Y, Qian Y, Sharif H. A survey on smart grid communication infrastructures: Motivations, requirements and challenges. IEEE Communications Surveys & Tutorials. 2012;**15**(1):5-20

[4] Wang K, Xu C, Zhang Y. Robust big data analytics for electricity price forecasting in the smart grid. IEEE Transactions on Big Data. 2017;**5**(1): 34-45

[5] Al-Dahidi S, Ayadi O, Alrbai M. Ensemble approach of optimized artificial neural networks for solar photovoltaic power prediction. IEEE Access. 2019;**7**:81741-81758

[6] Jihui Y, Craig F, Chikako A, Kazuo E. Predictive artificial neural network models to forecast the seasonal hourly electricity consumption for a University Campus. Sustainable Cities and Society. 2018;**42**:82-92

[7] Jihoon M, Jinwoong P, Eenjun H, Sanghoon J. Forecasting power consumption for higher educational institutions based on machine learning. The Journal of Supercomputing. 2018;**74**: 3778-3800

[8] Paul AA, Stephen A, Nkosinathi M, Obafemi OO. Hybrid adaptive neuro-fuzzy inference system (ANFIS) for a multi-campus university energy consumption forecast. International Journal of Ambient Energy. 2022;**43**:1685-1694

[9] Wang W, Lu Z. Cyber security in the smart grid: survey and challenges. Computer Networks. 2013;**57**: 1344-1371

[10] McGranaghan M, Schmitt DHL, Cleveland F, Lambert E. Enabling the integrated grid: leveraging data to integrate distributed resources and customers. IEEE Power and Energy Magazine. 2016;**14**:83-93

[11] Chen M, Shiwen M, Yunhao L. Big data: A survey. Mobile Networks and Applications. 2014;**19**:171-209

[12] Yan Y, Qian Y, Sharif H, Tipper D. A survey on smart grid communication infrastructures: motivations, requirements and challenges. In: IEEE Commun Surv Tutor. Vol. 15. NYC, USA: IEEE; 2013. pp. 5-20

[13] Siano P. Demand response and smart grids—a survey. Renewable and Sustainable Energy Reviews. 2014;**30**:461-478

[14] Fang B, Yin X, Tan Y, Li C, Gao Y, Cao Y, et al. The contributions of cloud technologies to smart grid. Renewable and Sustainable Energy Reviews. 2016; **59**:1326-1331

[15] Jaradat M, Jarrah M, Bousselham A, Jararweh Y, Al-Ayyoub M. The internet of energy: smart sensor networks and big data management for smart grid. Procedia Computer Science. 2015;**56**: 592-597

[16] Singla A, Sachdeva R. Review on security issues and attacks in wireless sensor networks. International Journal of

Advance Research in Computer Science Software Engineering. 2013;**3**:529-534

[17] Khushboo G, Vaishali S. Design issues and challenges in wireless sensor networks. International Journal of Computers and Applications. 2015: 112-126

[18] Malhotra J. Review on security issues and attacks in wireless sensor networks. International Journal of Future Genereration Communication Network. 2015;**8**:81-88

[19] Alidaee B, Wang H, Huang J, Sua LS. Integrating Statistical Simulation and Optimization for Redundancy Allocation in Smart Grid Infrastructure. Energies. 2023;**17**(1):255

[20] Attarha S, Narayan A, Hage HB, Krüger C, Castro F, Babazadeh D, et al. Virtualization management concept for flexible and fault-tolerant smart grid service provision. Energies. 2020;**13**(9): 2196

[21] Fengjun L, Luo B. Preserving data integrity for smart grid data aggregation. In: IEEE Third International Conference on Smart Grid Communications (SmartGridComm). 2012

[22] Shukla S, Thakur S, Breslin JG. Secure communication in smart meters using elliptic curve cryptography and digital signature algorithm. In: IEEE International Conference on Cyber Security and Resilience. 2021

[23] Vera BA, Colomo PR, Molloy O. Business process analytics using a big data approach. IT Professional. 2013;**15**: 29-35

[24] Stimmel CL. Big Data analytics strategies for the smart grid. Boca Raton: CRC Press; 2014

[25] Nga DV, See OH, Quang DN, Xuen CY, Chee LL. Visualization techniques in smart grid. Smart Grid Renewable Energy. 2012:3-175

[26] Lněnička M. Ahp model for the big data analytics platform selection. Acta Information Pragensia. 2015;**4**:108-121

[27] Shyam R, Ganesh HBB, Kumar SS, Prabaharan P, Soman KP. Apache spark a big data analytics platform for smart grid. Procedia Technology. 2015;**21**: 171-178

[28] Liu G, Zhu W, Saunders C, Gao F, Yu Y. Real-time complex event processing and analytics for smart grid. Procedia Computer Science. 2015;**61**:113-119

[29] Voyant C, Notton G, Kalogirou S, Nivet ML, Paoli C, Motte F, et al. Machine learning methods for solar radiation forecasting: A review. Renewable Energy. 2017;**105**:69-582

[30] Bourdeau M, Qiang ZX, Nefzaoui E, Guo X, Chatellier P. Modeling and forecasting building energy consumption: A review of data-driven techniques. Sustainable Cities and Society. 2019;**48**:101-533

[31] Foucquier A, Robert S, Suard F, Stéphan L, Jay A. State of the art in building modelling and energy performances prediction: A review. Renewable and Sustainable Energy Reviews. 2013;**23**:272-288

[32] Ordiano JAG, Bartschat A, Ludwig N, Braun E, Waczowicz S, Renkamp NP, et al. Concept and benchmark results for big data energy forecasting based on apache spark. Journal of Big Data. 2018;**5**:11

[33] Liu H. Smart cities: Big data prediction methods and applications. Berlin, Germany: Springer Nature; 2020

[34] Hyndman RJ, Athanasopoulos G. Forecasting: Principles and practice. Melbourne, Australia: OTexts; 2018

[35] Chambers B, Zaharia M. Spark: The definitive guide: Big data processing made simple. California, U.S: O'Reilly Media, Inc; 2018

[36] Sharma H, Kumar S. A survey on decision tree algorithms of classification in data mining. International Journal of Science and Research (IJSR). 2019;**5**: 2094-2097

[37] Kazemzadeh MR, Amjadian A, Amraee T. A hybrid data mining driven algorithm for long term electric peak load and energy demand forecasting. Energy. 2020;**204**:117-948

[38] Chen Y, Jia Z, Mercola D, Xie X. A gradient boosting algorithm for sur vival analysis via direct optimization of concordance index. Computational and Mathematical Methods in Medicine. 2013;**2013**:18-27

[39] Grabusts P, Zorins A. The influence of hidden neurons factor on neural network training quality assurance. In: Proceedings of the 10th International Scientific and Practical Conference. Latvia: Rezekne Academy of Technologies; 2015

[40] Duraisamy K, Zhang ZJ, Singh AP. New approaches in turbulence and transition modeling using data-driven techniques. In: 53rd AIAA Aerospace Sciences Meeting. VA, U.S: AIAA; 2015

[41] Parish EJ, Duraisamy K. A paradigm for data-driven predictive modeling using field inversion and machine learning. Journal of Computational Physics. 2016;**305**:758-774

[42] Amber KP, Aslam MW, Mahmood A, Kousar A, Yamin YM, Akbar B, et al. Energy consumption forecasting for university sector buildings. Energies. 2017;**10**:1579

[43] Grolinger K, Capretz MAM, Seewald L. Energy consumption prediction with big data: Balancing prediction accuracy and computational resources. In: IEEE International Congress on Big Data (BigData Congress). Washington, U.S; 2016. p. 2016

[44] Ruiz LGB, Rueda R, Cu, Pegalajar MC. Energy consumption forecasting based on Elman neural networks with evolutive optimization. Expert Systems with Applications. 2018;**92**:380-389

[45] Moon J, Park J, Hwang E, Jun S. Forecasting power consumption for higher educational institutions based on machine learning. The Journal of Supercomputing. 2018;**74**:3778-3800

[46] Allab Y, Pellegrino M, Guo X, Nefzaoui E, Kindinis A. Energy and comfort assessment in educational building: Case study in a French university campus. Energy and Buildings. 2017;**143**:202-2019

[47] Amber KP, Aslam MW, Hussain SK. Electricity consumption forecasting models for administration buildings of the UK higher education sector. Energy and Buildings. 2015;**90**:127-136

[48] Liang Y, Niu D, Wei CC. Short term load forecasting based on feature extraction and improved general regression neural network model. Energies. 2019;**166**:653-663

[49] Runge J, Zmeureanu R. Forecasting energy use in buildings using artificial neural networks: A review. Energies. 2019;**12**:3257

[50] Wei Y, Zhang X, Shi Y, Xia L, Pan S, Wu J, et al. A review of data-driven

approaches for prediction and classification of building energy consumption. Renewable and Sustainable Energy Reviews. 2018;**82**: 1027-1047

[51] Amasyali K, El-Gohary NM. A review of data-driven building energy consumption prediction studies. Renewable and Sustainable Energy Reviews. 2018;**81**:192-1205

[52] Amber KP, Ahmad R, Aslam MW, Kousar A, Usman M, Khan MS. Intelligent techniques for forecasting electricity consumption of buildings. Energy. 2018;**157**:886-893

[53] Kim MK, Kim YS, Srebric J. Predictions of electricity consumption in a campus building using occupant rates and weather elements with sensitivity analysis: Artificial neural network vs. linear regression. Sustainable Cities and Society. 2020;**62**:102-385

[54] Ahmad T, Chen H, Guo Y, Wang J. A comprehensive overview on the data driven and large scale based approaches for forecasting of building energy demand: A review. Energy and Buildings. 2018;**165**:301-320

[55] Ai S, Chakravorty A, Rong C. Household power demand prediction using evolutionary ensemble neural network pool with multiple network structures. Sensors. 2019;**19**:721

[56] Chammas M, Makhoul A, Demerjian J. An efficient data model for energy prediction using wireless sensors. Computers and Electrical Engineering. 2019;**76**:249-257

[57] Wahid F, Ghazali R, Shah AS, Fayaz M. Prediction of energy consumption in the buildings using multi-layer perceptron and random forest. IJAST. 2017;**101**:13-22

[58] Willmott CJ, Matsuura K. Advantages of the mean absolute error (MAE) over the root mean square error (RMSE) in assessing average model performance. Climate Research. 2005; **30**:79-82

[59] Kolassa S, Schutz W. Advantages of the MAD/MEAN ratio over the MAPE. Foresight: The International Journal of Applied Forecasting. 2007:40-43

[60] Lu Y, Wang W. Analysis of the mean absolute error (MAE) and the root mean square error (RMSE) in assessing rounding model. In: IOP conference series: materials science and engineering. 2018

[61] Chai T, Draxler RR. Root mean square error (RMSE) or mean absolute error (MAE)?–Arguments against avoiding RMSE in the literature. Geoscientific Model Development. 2014; 7:1247-1250

[62] Acharya MS, Armaan A, Antony AS. A comparison of regression models for prediction of graduate admissions. In: 2019 international conference on computational intelligence in data science (ICCIDS). 2019. pp. 1-5

[63] Zhang Y, Yang Q. A survey on multi-task learning. IEEE Transactions on Knowledge and Data Engineering. 2021;**34**(12):5586-5609

Chapter 3

ICT Project Field Report on the Development of a Local Smart Micro Grid in the Housing Industry

Steffen Späthe and Peter Conrad

Abstract

As part of our research project "WINNER Reloaded," we embarked on a three-year journey to oversee the establishment of a local micro smart grid in a residential community in the city of Chemnitz. Throughout this endeavor, we gained practical insights into the planning, implementation, and administration of the systems, acquiring a large amount of real-world energy data on the way. In the following chapter, we delve into the intricacies of our research project, offering a closer examination of its details and the wisdom we accrued through our experiences. Our intention is to provide a pragmatic guide that will facilitate others in embarking on the creation of their own local micro smart grid through judicious foresight and planning.

Keywords: micro smart grid, renewable energy, electric mobility, photovoltaics, landlord-to-tenant electricity, system architecture

1. Introduction

A local micro smart grid is a small-scale, intelligently managed energy distribution network designed to efficiently distribute electricity within a closed, specific local area, such as a residential community or a neighborhood. The objective of the Local Smart Micro Grid (LMSG) is to ensure electric power supply and optimize the supply and distribution of electrical energy within the local grid. Therefore, it incorporates local renewable energy sources like solar panels and often integrates energy storage systems. The key features to optimize operation include real-time monitoring and control of energy production and consumption and enabling dynamic adjustments to match supply with demand. Micro smart grids promote sustainable energy use, reduce grid stress, and enhance reliability, making them a valuable component of modern, reliable, and environmentally friendly power grids.

The benefits of a local micro smart grid come with significant planning, implementation, and operation efforts. Starting with defining requirements and developing an architecture, suitable technical components must be chosen. The focus should be not solely on meeting technical requirements but also on quality attributes like interoperability and the ability to integrate into a heterogeneous system. Building a complex system as an LMSG cannot be accomplished by a single party; among others, it involves partners with backgrounds in housing management, the electrical industry, building administration, and software development. Weighing mutual interests and continuous consensus-building is essential. Lastly, the diverse components require ongoing maintenance and support beyond the initial setup, which should also be considered.

During our three-year research project "WINNER Reloaded" [1], we participated in setting up a local micro smart grid. This endeavor allowed us to shape the direction of this system and accumulate a substantial corpus of real-world energy data, which we systematically analyzed throughout the project's duration.

In this comprehensive chapter, we share some practical experiences we gained and look into the dataset we accumulated. We aim to offer insights that go beyond the technical details of the LMSG itself.

The structure of our chapter is as follows: We start with an overview of our research project, highlighting its key objectives. Then, we delve into the technical and system architecture that underpins our LMSG, explaining how the system works in detail. In the next section, we explore the practical aspects of our project. Here, we share our accumulated experiences regarding the setup and development of system components and project administration and provide insights into the energy data we acquired throughout the project. Finally, we summarize the insights we gained and draw conclusions. Through this chapter, we hope to inspire new innovation and understanding in the realm of sustainable energy solutions.

2. Our research project "WINNER reloaded"

The research project "WINNER Reloaded" is situated within the context of promoting electromobility and establishing tenant electricity models. The project aims to realize demand-oriented charging infrastructure, tenant electricity supply, and decentralized energy generation for tenants, businesses, and other third-party users. Companies from the housing industry, the energy industry, and system installers are working together on this.

The main demonstrator, located within a new construction project in the city of Chemnitz/Germany, is at the heart of the research project. This complex consists of four residential buildings with 40 residential units, carports, and underground parking spaces. The parking spaces are equipped with charging infrastructure according to the tenants' needs. Notable features are a public high-performance charging station and the integration of a car-sharing station with electric vehicles as an additional mobility offer in the neighborhood.

Power supply for the tenants and the private and public charging infrastructure is sourced from solar power generated on-site using photovoltaic systems. To ensure the power supply, the residential complex has its dedicated network station connected to the local medium-voltage electricity grid. The project employs an intelligent micro smart grid, a small-scale dynamic power supply network.

The University of Jena manages the decentralized data collection in the local micro smart grid, which serves for load and charging management and forms the basis for electricity billing. Particular emphasis is placed on a high temporal resolution of the collected consumption data. The data collected this way is also made available to tenants for individual real-time monitoring. Within this system and with the data, we can also calculate and simulate dynamic pricing and billing models based on factors such as time of use and internally generated energy.

3. Architecture of our smart grid

The architecture of our smart grid has been based upon the requirements previously worked out in [2, 3]. It consists of a multitude of technical components that need to be independently integrated into the network and individually read and controlled. These components include multiple electricity meters, programmable logic controllers (PLCs) for capturing relevant energy flow parameters within the neighborhood, PV inverters, and electric vehicle charging stations. All components are connected *via* an internal local area network. A server within the internal network is available for data processing and control. An externally accessible system called the "Tenant Information Portal" exists for visualizing consumption data for the tenants.

An initial challenge in connecting such diverse components is achieving a unified integration into the overall infrastructure. Depending on the device and manufacturer, the methods for data retrieval differ significantly. In the case of our inverters and charging stations, data collection was achieved through REST requests; however, this method was not documented for either device and had to be determined through reverse engineering of their user interfaces. In contrast, the PLCs came with preinstalled software that inherently allowed data transmission at defined intervals *via* the Message Queuing Telemetry Transport (MQTT) protocol, although in a format predetermined by the manufacturer. To date, there has been no generally accepted IoT protocol-based standard to transmit energy measurements.

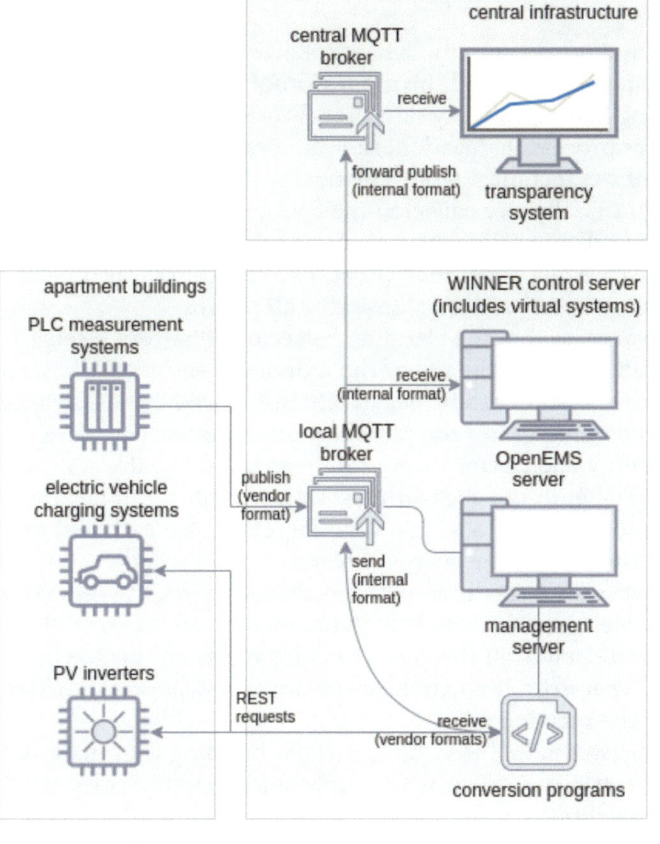

Figure 1.
An overview of the components in our local micro smart grid.

An important step in establishing a uniform infrastructure therefore is integrating heterogeneous components and selecting a common protocol and data format for the exchange of measurement and control data. In our case, we chose the MQTT protocol, designed for communication between Internet of Things (IoT) devices, through which measurement data was transmitted in an internally specified JSON-based format. To ensure data privacy and increase fault tolerance, the MQTT broker was installed on a local server within the LMSG. In this way, the raw data does not have to physically leave the local network infrastructure before it is processed further.

Data retrieval had to be performed individually for each component, necessitating the development of specialized software, also configured as services on the server. In total, there are three such components: two for data retrieval from the inverters and charging stations and one for processing and converting the measurement data received from the PLCs. Additionally, a customized version of the open-source energy management system OpenEMS [4] was deployed, serving for system status visualization and dynamic configuration. An overview of the software architecture is provided in **Figure 1**.

4. Purposes of the measurements and selected data points and sampling rates

In our local smart grid system, data is collected with two objectives. These objectives are the system control and the use for proof and transparency, among other things, for billing and plausibility purposes. From these two objectives, different requirements for processing location, sampling rate, and data storage are derived.

Awareness of the technical requirements for the data acquisition process developed iteratively. Initially, we collected the energy meter data only. The energy meter continuously records the active energy consumed, in the case of a household, or provided, in the case of a generation plant. The recording was done in the three-phase power supply as one total meter value across all phases.

With the awareness that the charging behavior of battery electric vehicles varies significantly with regard to the use of the individual electrical phases and that the overall local power system can be negatively influenced by unbalanced loads, we expanded the recording to include phase-accurate measured values.

Although energy and power values can be mathematically converted into each other, this conversion in real measuring systems is subject to narrow limits due to the sampling rate and data accuracy. We have therefore also gathered the instantaneous active power measured by the energy meters.

Finally, we became aware that in some constellations, the reduction of charging power by battery electric vehicles is accompanied by an increase of reactive power. In detail, this effect depends on the vehicle model and is not visible in active power measurements. Therefore, we extended the data acquisition by the phase-exact measured values of the reactive power.

The listed measurements were collected per building for tenant households and general home electricity, as well as for public and nonpublic charging infrastructure and generation facilities.

In addition to the measured values of the power grid, we directly connected to the charging infrastructure and the photovoltaic inverters and read out internal measured values and control parameters. As described, these systems do not provide an explicit software interface for integration. We created tools to access the required data *via* the existing web-based user interface. The creation of the tools was costly, and the implementation itself was always compromised by changes in the user interface due to software updates.

All measured values were autonomously and continuously determined by the measuring instruments or their independently running programs for interrogation. The sampling rate was uniformly set to 1 sample per 10 seconds for all measured values.

This temporal resolution appears to be very high and unnecessary for some physical quantities and use cases. On the other hand, short-term loads and fluctuations can only be seen with this high temporal resolution. In the context of the research project, there was no reason to reduce the sampling rate. We were not limited by processing speed, bandwidth, or memory in our quite small setup. For an optimized productive commercial system setup, the quantity and resolution of the measured values could be discussed again individually for each measurement.

As already discussed, after data acquisition, the measured values were converted into a uniform data format and made available to interested internal and external system components by means of a message middleware. Regardless of this, we stored all the data collected in an external time series database for later review and statistical analysis.

5. Selected data analyses

This section looks at some data analyses to underpin statements made previously. The selected analyses consider three different perspectives of the overall system that build on one another. The first is to analyze the distribution and predictability of generation and consumption. This analysis results in the degrees of freedom through load shifting for operational management. However, an effective operational management requires valid measured values. So, the second analysis highlights possible effects due to different temporal sampling rates and aggregation periods, accordingly. The necessity of phase-accurate measurement is illustrated in a third analysis using the example of the different charging behavior of vehicles.

To study the degrees of self-sufficiency and the degrees of freedom of load shifting, we analyze local production and consumption. In our two demonstration sites, we have two types of power generation sources, photovoltaics and combined heat and power generation. The combined heat and power plants are operated on a heat-led basis, with electricity generation being “only” a positive side effect. Since the cogeneration units in the project are not operated in the residential structure, we consider only the photovoltaic system in the following.

The energy generation of the photovoltaic system is highly dynamic and dependent on external influencing parameters. The generation output during the course of the day, with very low disturbing influences, for example by clouds, fog, or snow, is shown in **Figure 2**. The photovoltaic systems in our system are mounted on four houses and vary in the number of individual PV modules and orientation. For better comparability, the energy generated is therefore standardized to the nominal generation capacity (kWp). To illustrate the seasonal influence, power curves from winter (January), spring (April), and summer (June) are shown. The seasonal influence with

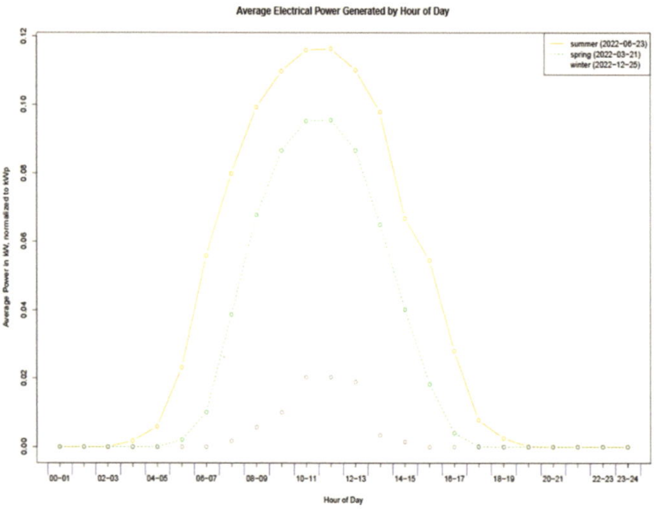

Figure 2.
Daily pattern of generation power on three exemplary days in summer, spring, and winter.

regard to maximum output and the start and end of generation according to the daily sunshine pattern is clearly recognizable. The shift of the generation period due to the different daylight times between summer and winter amounts to a maximum of approximately 8 hours.

The seasonal impact on the generation performance can also be seen very well in the overview of the annual course of the energy generated per day as shown in **Figure 3**. Compared to the peak values in summer (June, July), the plant at this location and in this fixed setup in winter (December, January) produces only approximately 5–10% of the energy.

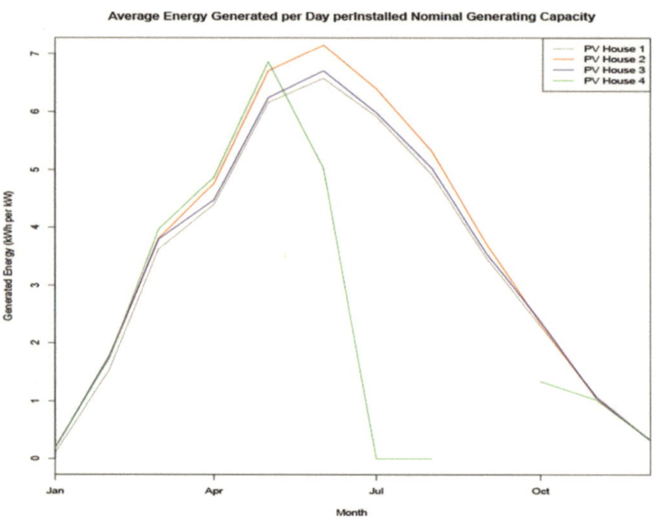

Figure 3.
Energy production in kWh per installed generation capacity (kWp) on a daily average per month of the four photovoltaic systems in the residential system design in the project.

Systematically, generation output of the photovoltaic system depends on unchangeable technical parameters as well as seasonal and weather influences.

The most important technical parameters are in particular

- the specific type of photovoltaic modules used,
- the geographical location of the system,
- the number and orientation of the photovoltaic modules in the specific system, and
- the course of the horizon, which affects the partial or complete shading of individual modules or the entire system.

The most important seasonal and weather-related parameters are

- position of the earth in relation to the sun,
- course of the sun's position throughout the day,
- degree and type of cloud cover throughout the day,
- rain or snowfall, both throughout the day and in the previous days,
- ambient temperature.

Based on these fixed technical parameters and the dynamic seasonal and weather-dependent parameters, the expected generation capacity for the next day can be predicted. For this purpose, there are modeling approaches that attempt to represent the system completely as a physical system. There are also forecasting models that are based on statistical methods or machine learning methods.

We experimented with a hybrid model in our project. We described the photovoltaic system physically and used this model to calculate the theoretical generation output without interfering weather influences. Based on these theoretical maximum values, we used external weather forecasts and statistical models to derive a day ahead forecast that takes the weather into account.

Whether purely physical, statistical, or hybrid, all forecasting models can provide sufficiently accurate planning values that can be used in a day-ahead planning. The selection of the appropriate model depends heavily on the available initial data. If no constructive technical information is available, physical modeling will not be successful. The same applies to statistical models if no real measured values are available. It should be noted that regardless of the quality of the forecast model, intraday real-time control is always absolutely essential, as the actual generation and consumption output always deviates from the planned values.

Unlike PV-based generation, energy use by tenants is less easy to model and predict. For this task, primarily only statistical forecast methods are available. Total energy consumption can be disaggregated into different energy consumption categories. Some main categories are

- Household operations, entertainment, and light,

- Preparation and provision of hot water,
- Provision of heating and air conditioning, and
- Battery electric vehicle (BEV) charging.

Alongside the energy consumed by the residents, there is also the energy required for the general operation of the building infrastructure, for instance, house lighting, emergency lighting, elevators, and ventilation. Each of these energy consumption categories has its own individual characteristic consumption time profile. The energy consumption for an apartment is most strongly influenced by the question of whether this apartment is currently occupied. This aspect seems trivial at first glance, but it is not only determined by the general occupancy status of the apartment, which could be retrieved from an external management system. In particular, this factor must be taken into account during typical vacation periods and holidays, when apartments are regularly unoccupied.

If the apartment is occupied, the energy consumption in the household is strongly influenced by the classification of the day. The days can be divided into the classes "Weekday," "Saturday," and "Sunday/Holiday". The season naturally influences energy consumption in the household, as well. The average energy profiles for households measured in the project are shown in **Figure 4**.

Figure 5 shows that the time of year also has an influence on the energy consumption of households. In the winter months, energy consumption by households is visibly higher than in the summer months. The high energy consumption in December is remarkable. This is due to the Christian holidays and the associated traditions of increased baking and cooking. The energy consumption in summer, on the other hand, is noticeably lower. This is partly due to the longer daylight hours and therefore

Figure 4.
Daily pattern of average electrical power drawn by households dependening on day of the week and season.

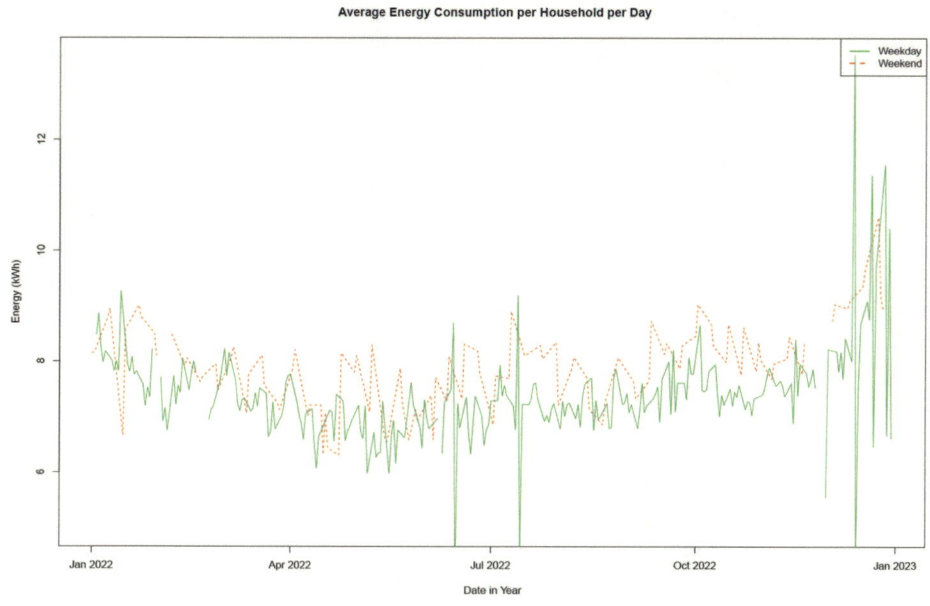

Figure 5.
Average electrical power consumption by households on weekday and weekend.

less need for artificial light and more activities outside the household, as well as the temporary absence of residents during their summer vacation.

As previously mentioned, it is absolutely essential to measure both the curves of the energy meter and the instantaneous power.

Depending on their usage, the energy meters only provide measured values of one order of magnitude—kWh. With a sampling rate of 1 measured value per 10 seconds, this order of magnitude allows very high power levels to be measured with sufficient accuracy. These are, for example, charging processes for electric vehicles. However, the typical standby power of the charging infrastructure is 1500 times lower and cannot be determined with sufficient accuracy by calculating the difference between two consecutive meter values of the energy meters (maximum charging power approx. 150 kW, standby power approx. 100 W). This finding can be clearly seen in **Figure 6** in the time window between 12:30 and 12:45. The curve based on 15 min measured values shows no value. The curve based on the difference of 30 sec values shows a zero curve with individual spikes. These spikes result from the "jumping" of the read-out measured value according to the smallest possible interval.

One disadvantage of recording the amount of energy per unit of time compared to recording meter values for the total energy is the additional effort required to map alternative time series. For example, if an amount of energy per 5 min is determined and stored, the determination of time series with resolutions of 1 min, 60 min, or 24 h is associated with additional effort.

The aggregated recording of measured values across all phases, especially power values, has already been mentioned as unfavorable. Phase-specific measurement and, if possible, control proved to be more sensible.

Figure 7 shows the difference between the aggregated power measurement and the phase-accurate power measurement. The figure shows the three different charging processes as they were actually measured. The first charging process is a three-phase charging process with a total power of 11 kW. Accordingly, all phases L1, L2,

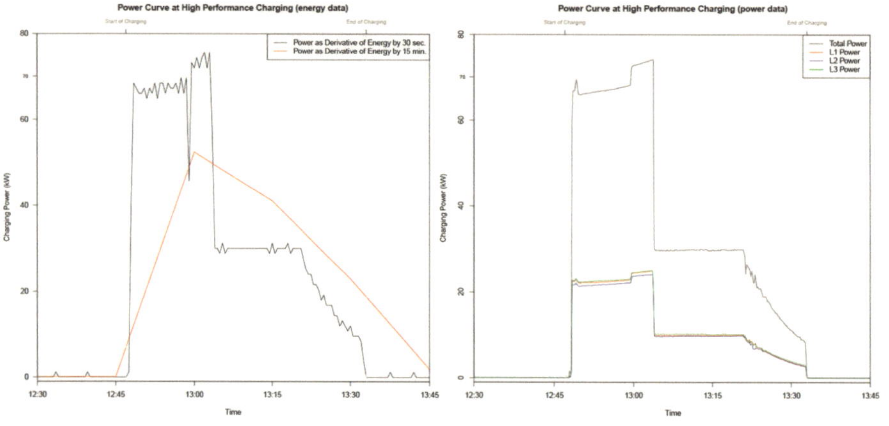

Figure 6.
Electrical power used by high-performance charging.

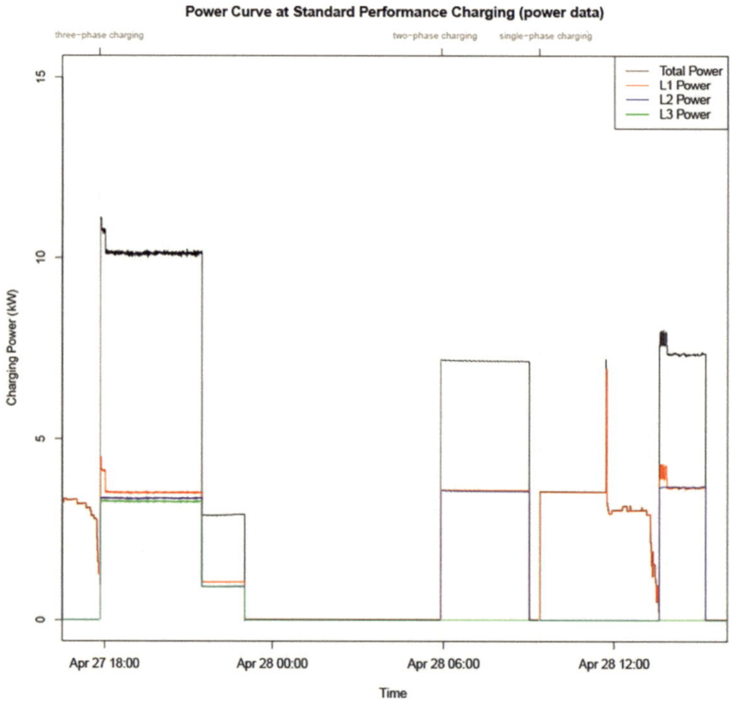

Figure 7.
Electrical power used by standard charging in different charging modes.

and L3 are utilized almost equally. The second charging process is only a two-phase process with a total output of 7.2 kW, whereby phase L3 is not utilized. The third charging process only utilizes phase L1 and has a charging power of 3.6 kW.

However, in our setup, the control of the photovoltaic inverters and the charging infrastructure made it impossible to react to load asymmetries and take appropriate countermeasures. Future systems have to pay more attention to the necessity of balancing asymmetrical loads in the local grid and must include appropriate technical options for counteracting them.

6. Our experiences in setting up and operating a local micro smart grid

During the setup phase, a significant hurdle encountered was the integration of individual hardware components, as it demanded substantial analytical efforts and subsequent software development time for automated data access. Similarly, the operation of charging stations for the purpose of dynamic power regulation proved difficult due to the lack of a unified interface and a general high inertia of the system. At this point, we became aware that components readily available on the market may not necessarily be well-suited for integration into a smart grid setup, as off-the-shelf products do not necessarily provide robust features for automatic data access or dynamic control.

The selection of suitable components for a local micro smart grid necessitates a meticulous definition of prerequisites beforehand, which should be thoroughly scrutinized before procurement, if necessary in consultation with the manufacturer. To circumvent additional work, we strongly advise that this step be undertaken with great precision during the planning phase and that requirements be harmonized with the project partners responsible for provisioning (typically possessing extensive experience in the construction of conventional systems).

During the operation of the system, additional challenges related to data measurement emerged. Notably, the need for configuring a central time synchronization service became apparent to correlate timestamps of measurement data, thereby compensating for and mitigating potentially significant discrepancies that arose over extended periods, varying by device type. This measure was essential to avert potential control errors stemming from these time disparities.

As previously described, we realized the need to record phase individual measurement values rather than examining them solely in aggregates. This is of particular importance since high power applications, like charging EV or cooking, can lead to highly asymmetrical loads on individual electrical phases. Such imbalances between the phases are not recognizable in aggregates, but have to be identified. Asymmetrical loads lead to a higher load on the neutral conductor, for which it may not be designed. Consequently, it is advisable to design the data model and processing infrastructure during the planning phase with consideration for this larger volume of data. Retroactive adjustments can cause considerable work and expense.

Regarding the issue of unbalanced loads caused by electric vehicles, this is recognized in the electrical industry and various strategies have been developed to solve it. Single-phase charging electric vehicles tend to predominantly draw power from pin 1, while two-phase charging vehicles typically draw power from pin 1 and 2.

One pragmatic approach is therefore to rotate phases over several subsequent charging points. In this way, pin 1 of the charging cable is connected to phase L1 at the first charging point, to phase L2 at the next charging point, and to phase L3 at the third charging point. This straightforward approach enables a more even distribution of loads, even if only single-phase charging vehicles are used. However, this wiring method introduces the disadvantage that measured values within the measurement installation can no longer be easily attributed to the correct phase or pin. This is because the "phase L1" recorded at a charging point, in reality, could be connected to phase L2 or phase L3 within the overarching network. These intricacies must be clarified with the partner responsible for the electrical installation to prevent erroneous measurements and subsequent incorrect control operations. Beyond the case of the charging infrastructure, the consistent wiring of the electrical phase is generally an extremely important factor. Swapping or rotating the three phases within the distributed electrical system will lead to incorrect measured values and subsequently to implausible control behavior. Finding this cause is extremely difficult and time-consuming.

Finally, we would like to emphasize that a local micro smart grid is a complex system comprising numerous interdependent hardware and software components that require regular maintenance. Component failures can occur at any time and should be promptly identified and addressed. Examples of problems we encountered during the project include:

- Crashes of software due to various causes (e.g., lack of memory).
- Solar inverter hardware failures and the resulting loss of PV energy.
- Issues in the hardware of the local communication infrastructure and a resulting loss of interconnectivity and subsequently measurement data.
- Non-static system configurations caused changes, rendering devices unreachable for data exchange; for instance, queried inverters changed IP address following a reboot.
- Setup changes caused by non-static system configurations which renders devices unreachable for data exchange (e.g., queried inverters IP addresses changed following a reboot).

The variety of these errors should make it clear that LMSGs are not self-perpetuating and that appropriate monitoring of the operation should be factored in during the entire operation, beyond the initial planning and building phase. We would definitely recommend designating a special system manager for this purpose.

The large number of fallacies that we encountered in the course of the project prompted us to systematically record these in a separate article and address them in more detail [5].

7. Conclusion

The "WINNER Reloaded" project afforded us the opportunity to construct a local micro smart grid from the ground up and, in the process, accumulate a substantial amount of new experiences and data. Over the course of 3 years, we gained a profound insight into the multifaceted practical aspects of such a system. From these experiences, we aim to emphasize the divergence from conventional IT systems and emphasize the importance of proactive planning and competent administration.

Moreover, we aspire that our data analyses have provided a glimpse into load behavior across different times and situations, potentially bearing relevance for future local micro smart grid planning. We hope that our chapter will contribute to paving the way for a sustainable future.

Acknowledgements

We would like to thank all project partners of the "WINNER Reloaded" project. Without their visions, work on the project, and technical expertise in its implementation, the project and this report would not have been possible.

Conflict of interest

The authors declare no conflict of interest.

Thanks

The research project "WINNER Reloaded" was funded by the Federal Ministry for Economic Affairs and Energy of Germany under project number 01ME19001E.

Author details

Steffen Späthe*† and Peter Conrad†
Friedrich Schiller University Jena, Jena, Germany

*Address all correspondence to: steffen.spaethe@uni-jena.de

† These authors contributed equally.

References

[1] Chemnitzer Siedlungsgemeinschaft eG. WINNER reloaded Projekt. 2020. Available from: http://www.winner-projekt.de. [Accessed: Sep. 11, 2023]

[2] Späthe S, Conrad P. General requirements and system architecture in local micro smart grid. In: Krieger UR, Eichler G, Erfurth C, Fahrnberger G, editors. Innovations for Community Services. Cham: Springer International Publishing; 2021. pp. 239-250. ISBN 978-3-030-75004-6

[3] Hertrampf F, Apel S, Späthe S. Requirements and architecture concept for a data processing system in local area smart grid. Journal of Electrical Engineering. 2018;**6**:1-10. DOI: 10.17265/2328-2223. ISSN 2328-2223

[4] OpenEMS Association e.V. OpenEMS – The 100% energy revolution needs a free and open source. Energy Management System. 2022. DOI: 10.5281/zenodo.10449716. Available from: https://openems.io/

[5] Späthe S, Conrad P. Fallacies in the implementation of a local micro smart grid. In: 2022 International Conference on Computational Science and Computational Intelligence (CSCI), Las Vegas, NV, USA. IEEE; 2022. pp. 1241-1247. DOI: 10.1109/CSCI58124.2022.00224

Chapter 4

Advantages and Challenges of Using Renewable Sources with Smart Meters for Better Energy Efficiency-Use Case of a Brazilian Rural Area

Effrain Roney Bernardes

Abstract

With the innovation of electrical networks to serve large cities, the concept of smart city began to be disseminated to improve energy efficiency and the integration of renewable sources in their matrix, therefore, Information and Communication Technology (ICT) plays a role central in the evolution of electrical networks to smart grids. Generally, ICT plays a crucial role in facilitating the control, management, and comprehensive monitoring of operations within smart grids. The inherent complexity of smart grids arises from the intricate interplay of various heterogeneous components, diverse types of generators, and a multitude of load variations. Despite these challenges, the integration of advanced ICT solutions continues to pave the way for enhanced efficiency and reliability in smart grid operations. This technology enables real-time data collection, communication between devices, integration of renewable energy sources, and automation of the electrical grid. By facilitating advanced monitoring, control, and data analysis, ICT makes energy distribution more efficient, reliable, and sustainable, promoting the transition to a modern and smart electricity sector. In this chapter, we will address the advantages and disadvantages of renewable sources and the crucial role of ICT in this energy transformation to control and monitor smart grids to avoid the flow of power in utility networks.

Keywords: smart meter, smart grid, renewable energy, energy efficiency, distributed generation

1. Introduction

In the ever-evolving landscape of energy production and consumption, the integration of renewable sources and smart metering technologies stands as a pivotal frontier in the pursuit of sustainable and efficient energy systems. This book delves into the multifaceted realm of energy management, exploring the advantages and disadvantages inherent in the harmonious marriage of renewable sources and smart meters. As the global community grapples with the imperative to transition toward greener energy alternatives, understanding the intricacies of this transformative synergy becomes paramount. The utilization of renewable energy sources promises a cleaner, more sustainable future, offering the potential to reduce our dependence on finite fossil fuels. Concurrently, smart meters have emerged as indispensable tools, enabling real-time monitoring, data analysis, and adaptive control mechanisms to optimize energy usage. However, this amalgamation is not without its challenges. The book meticulously examines the complexities associated with integrating renewable sources with smart meters, addressing potential drawbacks and proposing innovative solutions to enhance energy efficiency while mitigating the risk of power flow inversion. Join us on a journey through the dynamic landscape of renewable energy and smart metering, as we navigate the advantages and disadvantages inherent in this symbiotic relationship. By shedding light on the intricacies of these technologies, we aim to contribute to a deeper understanding of how to harness their potential for a more sustainable and resilient energy future.

Then comes the concept of smart grids (SG), first cited in 2005 in an article entitled "Towards a Smart Grid," written by S. Massoud Amin and Bruce F. Wollenberg. Since then, several definitions have emerged for the concept of smart grids. However, the core of its definition is always associated with the use of information and sensing technology applied in a large-scale, interconnected, bidirectional, sustainable network infrastructure that implements agility and security in search of being resilient against threats and conditions not foreseen in the transportation of energy [1].

Presently, numerous countries globally find themselves compelled to reassess their investments in electricity infrastructure and overall energy generation, transmission, supply, and management [2]. This urgency has prompted utility companies worldwide to channel significant financial resources, totaling billions of dollars, into the revitalization and modernization of electricity grids through the incorporation of smarter technologies. The overarching goal is to enhance efficiency, reliability, privacy, and security in the realm of electricity usage, particularly as societies navigate the challenges posed by the digital age [3].

The envisioned future energy supply system revolves around the utilization of advanced and intelligent technologies to streamline the network control of distribution grids. In addition, innovative features aim to encourage customers to evaluate the type of generation prior to consumption, fostering a more mindful approach to energy utilization. Concurrently, electricity providers stand to capitalize on growth opportunities. The realization of these functionalities is anticipated through the implementation of a smart grid—an intelligent energy supply system. The term "smart grid" refers to the sophisticated networking and control of intelligent generators, storage facilities, loads, and network operating equipment within power transmission and distribution networks, facilitated by Information and Communication Technologies (ICT) [4].

2. What are smart meters?

A smart meter is characterized as an electric meter that records energy consumption data at defined intervals, possesses two-way communication capabilities, stores metering data in registers, and accommodates various tariffs such as time of use, inclined block, maximum demand, and free basic electricity. These meters can be remotely updated, have the capability to switch attached loads on command, and interface with data concentrators [5].

3. How to avoid power flow inversion?

In a traditional power grid, electricity flows from large generators through transmission lines to distribution networks and ultimately to consumers. Power Flow Inversion: This occurs when power flows in the reverse direction, from distribution networks back toward transmission lines or generators. It can happen due to high penetration of distributed energy resources (DERs) like solar panels and wind turbines. Power flow determines the behavior of electrical quantities (voltages, currents, and powers) throughout the circuits of the electrical system. To find out whether variations in these quantities will not degrade the stability of the network, a study of the power flow is carried out. The power flow study aims to predict the behavior of a system's electrical variables and ensure that all parameters are within the supportability limits of the equipment that makes up the network [6].

Why it's avoided:

- Grid instability: inversion can create voltage and frequency fluctuations, potentially leading to outages or equipment damage.
- Protection challenges: traditional protection devices are not designed to handle reverse power flows, potentially causing misoperation or delays in fault isolation.
- Wear and tear: inverters in DERs can experience increased stress and wear from sustained backflow of power.
- Low load conditions (when demand is low).
- Faults or disturbances in the grid.

4. Integration of renewable sources

4.1 Benefits

Clean and sustainable energy represents a paradigm shift in our approach to power generation, emphasizing environmental responsibility and long-term resilience. At its core, clean energy harnesses renewable sources like solar, wind, and hydropower, steering away from traditional fossil fuels that contribute to pollution and climate change. The adoption of these eco-friendly alternatives underscores a commitment to reducing carbon footprints and preserving natural ecosystems. This transition also leverages technological advancements, from highly efficient solar panels to innovative wind turbines, driving the evolution toward a greener energy landscape. **Figure 1** shows methodology of it.

Moreover, the embrace of clean energy is not merely a choice for today but a strategic move for tomorrow, tapping into infinite and diversified resources that ensure energy security and independence from geopolitically sensitive commodities. As the world confronts the challenges of climate change, clean and sustainable energy

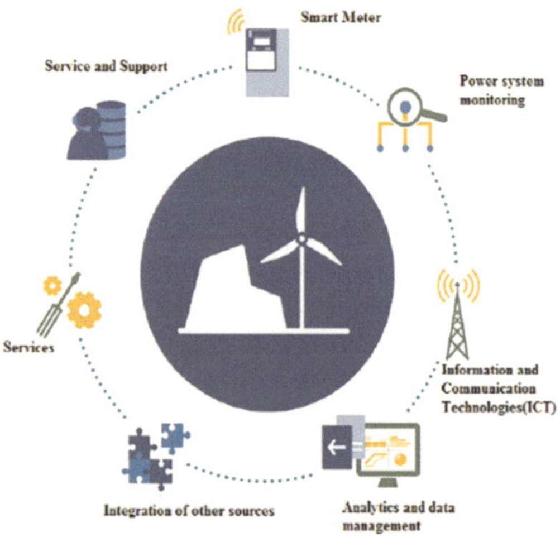

Figure 1.
Smart grid components. Source: Ref. [7].

emerges as a beacon of hope, fostering a harmonious coexistence between technological progress and environmental preservation. These networks can provide economic, social, and environmental benefits, but there are numerous challenges to be overcome, such as coexistence with conventional networks, real-time data traffic performance, treatment applied to the large volume of data acquired, software implementations, and specific hardware development, in addition to the vulnerabilities of information systems among others [8].

Diversification of energy sources is a strategic imperative in the pursuit of a resilient and sustainable energy landscape. This approach involves moving beyond traditional reliance on a single predominant energy source, such as fossil fuels, to integrate a mix of renewable and non-renewable sources. The benefits of diversification are multifaceted; it enhances energy security by mitigating the impact of supply disruptions and price fluctuations in any single resource. Moreover, incorporating renewable sources like solar, wind, and hydropower contributes to environmental sustainability by reducing greenhouse gas emissions and dependence on finite resources. This shift toward a diversified energy portfolio not only fosters innovation but also positions societies to adapt more effectively to evolving energy demands and global dynamics. Embracing a varied mix of energy sources stands as a cornerstone for a robust, flexible, and forward-looking energy infrastructure that can navigate the complexities of a rapidly changing world. Incorporating renewable energy into the electrical power grid presents numerous advantages for the power sector as well as the social, economic, and environmental domains. Concerning the environment, the electricity sector presently constitutes a notable source of carbon dioxide emissions [9].

Decentralized power generation marks a transformative departure from the traditional centralized model, where energy production is concentrated in large-scale facilities. In this innovative paradigm, power is generated closer to the point of consumption, often utilizing local renewable resources. Technologies such as solar panels, small-scale wind turbines, and microgrids empower communities to produce their own energy, fostering energy independence and resilience. Decentralization not only reduces transmission losses associated with long-distance power transport but also enhances the overall efficiency and reliability of the energy grid.

Figure 2.
Inverter applies at plant for doing grid zero. Source: Author himself.

This approach aligns with the principles of sustainability by minimizing environmental impact and increasing the integration of clean energy sources. As we witness a shift toward decentralized power generation, communities are becoming active contributors to their energy needs, promoting a more resilient and adaptable energy infrastructure for the future. The diverse changes signify a paradigmatic shift within the energy sector. The traditional model of one-way electricity distribution, flowing from generators to consumers with a consistent decrease in voltages, is being supplanted by a decentralized integration of renewable energies with an n:m relationship. This transition results in consumers assuming the role of producers, referred to as "Prosumers," and distribution models evolving from centralized to demand-driven, transforming formerly stable processes into dynamic ones. These alterations profoundly impact the involved stakeholders and necessitate a reevaluation of data and information management. The central challenges in managing information, especially with Smart Metering, are detailed in **Figure 2** across all levels of information management [10].

4.2 Challenges

The intermittent and variable nature of renewable energy sources, such as solar and wind power, poses both challenges and opportunities in the transition toward sustainable energy. Intermittency refers to the irregular availability of energy, often dictated by factors like weather conditions and time of day. Variability encompasses the fluctuations in energy output that these sources exhibit. While the unpredictability of renewable energy production can strain the stability of the power grid, advancements in technology, coupled with the integration of smart grid solutions, offer promising avenues to address these issues. Energy storage systems, demand-response strategies facilitated by smart meters, and enhanced forecasting algorithms are pivotal in mitigating the impacts of intermittency and variability. By effectively managing and adapting to the dynamic nature of renewable sources, we can harness their substantial benefits while ensuring a reliable and resilient energy supply for the future.

These networks can provide economic, social, and environmental benefits, but there are numerous challenges to be overcome, such as coexistence with conventional networks, real-time data traffic performance, treatment applied to the large volume of data acquired, software implementations, and specific hardware development, in addition to the vulnerabilities of information systems among others [11]. Information and Communication Technology (ICT) serves multiple functions within smart grids. Overall, ICT contributes to controlling, managing, and overseeing the entire operation of smart grids. The complexity of smart grids arises from their expansive structures, featuring a multitude of diverse components, various types of generators, and a range of load characteristics. The intricate interdependence among these components means that issues in one part can have ripple effects, potentially affecting other components and, in some cases, the entire grid [12].

The upfront costs of implementing renewable energy systems and smart metering infrastructure can be substantial. However, long-term benefits, such as reduced operational costs and environmental impact, often outweigh the initial investment. An item focusing on the initial capital investment required for the integration of renewable energy sources provides a comprehensive exploration of the financial aspects associated with adopting sustainable energy solutions. It delves into the various components that contribute to the upfront costs, elucidating the considerations involved in selecting specific renewable technologies such as solar, wind, or geothermal. The item elucidates the significance of site assessments, emphasizing the need for thorough studies to evaluate resource potential and address environmental concerns.

It sheds light on the procurement and installation of essential equipment, encompassing solar panels, wind turbines, and infrastructure necessary for grid connection or energy storage. Furthermore, the item discusses the critical role of engineering and design in optimizing system efficiency and compliance with regulatory standards. It recognizes the importance of navigating through permitting processes and accounting for land acquisition costs, underscoring the multifaceted nature of the initial investment. By exploring labor expenses and emphasizing the inclusion of contingency budgets to mitigate risks, the item provides valuable insights into the financial planning required for successful renewable energy integration.

5. Smart meters in energy management

5.1 Advantages

Smart meters play a pivotal role in revolutionizing energy management through real-time monitoring and control, offering numerous advantages in the realm of efficiency, sustainability, and consumer empowerment. Unlike traditional utility meters, smart meters provide real-time data on energy consumption, enabling both consumers and utility companies to make informed decisions. One notable advantage is the ability to track energy usage patterns, allowing consumers to identify peak demand periods and adjust their behavior accordingly to optimize energy efficiency. For utility companies, smart meters facilitate better grid management by providing instantaneous insights into power distribution and demand fluctuations. This real-time monitoring enhances the overall reliability and resilience of the electrical grid.

Additionally, smart meters enable the implementation of dynamic pricing models, encouraging consumers to shift their energy consumption to off-peak hours, contributing to load balancing, and reducing overall energy costs. The two-way communication capabilities of smart meters empower consumers to actively participate in demand-response programs, fostering a more interactive and sustainable energy ecosystem. In essence, smart meters serve as a cornerstone in modern energy management systems, offering a pathway toward a more efficient, responsive, and environmentally conscious energy infrastructure. ICT has several roles in smart grids.

Hence, the identification of faults in a specific grid section and promptly communicating them to other interconnected sections hold significant importance for ensuring the seamless operation of smart grids overall. In these aspects, Information and Communication Technology (ICT) plays pivotal roles, with the additional capability of autonomously addressing various issues [13].

Smart meters play a crucial role in facilitating demand-response initiatives within energy management systems, offering several advantages that contribute to a more flexible and responsive grid. One key advantage lies in the ability of smart meters to provide real-time data on energy consumption patterns. This enables utilities to identify peak demand periods and implement demand-response programs to curtail energy usage during these times. Through the two-way communication capabilities of smart meters, utilities can remotely and proactively communicate with consumers, encouraging them to reduce their electricity consumption during high-demand periods. This not only helps in preventing grid overloads but also minimizes the need for additional power generation, reducing stress on the overall energy infrastructure. The smart meter serves a dual function, acting as both a monitoring and control tool. It gathers data on the energy consumption from all areas continuously and updates this information to the utility, facilitating load demand forecasting [14].

Moreover, smart meters enable the implementation of dynamic pricing models, allowing utilities to adjust electricity rates based on demand fluctuations. This incentivizes consumers to shift their energy consumption to off-peak hours when electricity costs are lower, contributing to load balancing and optimizing the utilization of available resources. In essence, the integration of smart meters in demand-response programs enhances the overall efficiency and reliability of the electrical grid while empowering consumers to actively participate in sustainable energy practices. In Europe, significant industrial clients are enticed to participate in demand-response (DR) initiatives through the implementation of time-of-use pricing and interruptible tariffs. Additionally, certain industries employ substantial load avoidance criteria as a means to qualify for incentives through DR [15].

5.2 Challenges

While smart meters offer numerous benefits in energy management, their deployment raises valid concerns about privacy and security. One significant disadvantage is the potential invasion of privacy due to the granular data collected by smart meters. These devices provide detailed information about a household's energy consumption patterns, daily routines, and even specific appliance usage. The aggregation of such data raises concerns about the potential misuse or unauthorized access, posing a threat to individuals' privacy. Additionally, smart meters operate within a connected network, making them susceptible to cybersecurity threats. Unauthorized access to the metering data can lead to privacy breaches, exposing sensitive information about residents.

There is also the risk of malicious attacks on the grid infrastructure, potentially disrupting the reliability of the entire energy system. To address these concerns, stringent cybersecurity measures must be implemented to safeguard the data transmitted by smart meters. Moreover, clear and robust privacy policies and regulations are essential to ensure that consumer data is handled responsibly and ethically, mitigating the potential downsides associated with the deployment of smart meters in residential and commercial settings. Balancing the advantages of smart meters with robust privacy and security measures is crucial for fostering public trust and ensuring the responsible implementation of these technologies. Encryption is an often advocated measure for data security and privacy, but it is not effective against several classes of attacks in industrial control systems, and it can severely decrease visibility and monitorability of smart grid networks [14].

The deployment of smart meters in energy management systems presents inherent challenges, particularly in the realms of privacy and integration. One notable disadvantage is the potential compromise of privacy due to the detailed and real-time data collection capabilities of smart meters. The granularity of information renewable energy sources (RES) integrated into smart grids offer advantages, including enhanced peak demand management that contributes to the overall reliability and efficiency of the smart grid (SG). The SG achieves greater efficiency and dependability with a reduction in power quality issues and voltage regulation challenges. However, in smart grid applications, challenges arise concerning security, safety, and reliability. These challenges can be addressed through the implementation of a high-quality subsystem and advanced communication technologies. Regarding residents' energy consumption patterns can lead to concerns about the intrusive monitoring of individuals' daily activities and routines [16]. Apart from the utility and consumers, there is a risk of unauthorized data exposure, potentially allowing illegal entities to identify the population of a specific locality. Smart Meters (SM), along with communication technologies, offer the option of exchanging information with other SMs. This becomes problematic when consumers are uninterested in authenticating their energy information within their neighborhood. Smart grids, being connected to various operating conditions and protocols through the Internet, are susceptible to cyber-attacks, leading to serious issues such as damage and failure of smart grid subsystems [17].

Balancing the benefits of improved energy management with the protection of personal privacy is a delicate challenge. Additionally, the integration of smart meters into existing infrastructure poses technical challenges. Compatibility issues may arise when trying to integrate these advanced metering systems with legacy grid components and communication protocols. Retrofitting existing systems to accommodate smart meters can be complex and may require significant investments in upgrading infrastructure. Moreover, the diverse communication technologies employed by smart meters can lead to interoperability challenges, hindering seamless integration into a cohesive energy management network. Addressing these challenges requires careful consideration of privacy protection measures and systematic integration planning to ensure the effective deployment of smart meters without compromising individual privacy or disrupting the stability of existing energy infrastructure.

6. Achieving synergy: overcoming challenges

In the pursuit of achieving synergy within organizational frameworks, the role of Information and Communication Technology (ICT) is pivotal, albeit not without its challenges. ICT serves as a cornerstone in fostering collaboration, providing the necessary tools and platforms for seamless communication and information exchange. However, challenges such as technological disparities, data security concerns, and resistance to technological adoption can impede the effective integration of ICT into collaborative processes. Addressing these challenges requires strategic planning, investment in up-to-date technologies, and comprehensive training programs to ensure that teams can leverage ICT tools to their full potential. Moreover, as organizations increasingly rely on digital platforms for collaborative work, ensuring the resilience and cybersecurity of ICT infrastructure becomes paramount. Overcoming these challenges allows organizations to harness the power of ICT, enhancing communication, promoting collaboration, and ultimately contributing to the successful attainment of synergy in a dynamic and interconnected business environment [13].

Due to security concerns, consumers risk compromising their privacy and safety when using public cloud services. Preserving confidentiality in data becomes a crucial challenge in deploying cloud computing for smart grids (SG). Individual data, such as energy consumption and home appliance usage, needs secure storage and management. In ensuring cloud confidentiality, attention should be directed toward the consistency and secure transfer of data. Cloud computing stores data in various locations, each maintained and controlled by the respective business or society [18].

7. Study of case

This case study explores the implementation of a grid zero system in a farm setting, **Figure 3**, leveraging the capabilities of three Growatt MIN10000TL-X inverters. The objective is to achieve energy independence and sustainability by integrating renewable energy sources with smart metering technology. The study delves into the technical aspects, challenges, and benefits of this innovative solution. Subsequently, we shall demonstrate how technology with grid zero and its advantages can be demonstrated.

Figure 3.
Project site situation plan. Source: Google Earth.

7.1 Objectives

As the demand for sustainable energy solutions rises, the agricultural sector is increasingly turning toward renewable energy sources to meet its power needs. This case study focuses on the deployment of three Growatt MIN10000TL-X inverters in a farm setting, aiming to achieve grid zero status through the integration of solar energy and smart metering.

The primary objectives of this case study are to:

a. Analyze the technical specifications of Growatt MIN10000TL-X inverters.

b. Investigate the installation process and challenges faced.

c. Evaluate the impact of integrating smart meters for efficient energy management.

d. Assess the overall performance and benefits of the grid zero system.

7.2 Technical specifications of Growatt MIN10000TL-X inverters

7.2.1 Inverter features

A detailed examination of the technical features of Growatt MIN10000TL-X inverters, highlighting their suitability for agricultural applications, is provided. In **Figure 3**, the installation of inverters is shown, at picture show three inverters power 10 kW each totally 30 kW. Capacity/Power Rating: The "MIN10000" in the model name suggests a capacity or power rating of 10,000 W or 10 kW.

- Type of inverter: it could be a grid-tied inverter, which is designed to synchronize with the utility grid.

- Efficiency: look for information on the inverter's efficiency, which indicates how effectively it converts DC power from solar panels to AC power for use in your home or to feed back into the grid.

- Maximum Power Point Tracking (MPPT): many inverters use MPPT technology to optimize the power output from solar panels by adjusting the operating point.

- LCD display/communication: a user-friendly display or interface that provides information about the system's performance, including power output, energy production, and potential issues.

- Monitoring and communication options: some inverters offer monitoring features that allow you to track your system's performance remotely. This could include WiFi or Ethernet connectivity.

- Protection features: inverters often come with built-in protection mechanisms such as overvoltage protection, overcurrent protection, and anti-islanding protection.

- Cooling system: inverters may have a cooling system, such as fans or heatsinks, to prevent overheating during operation.

- Compliance with standards: check whether the inverter complies with relevant industry standards and certifications.

- Warranty: inverter warranties typically cover a certain number of years. Verify the warranty period and what it covers.

7.3 Installation process and challenges

7.3.1 System configuration

Step-by-step description of the installation process, emphasizing the integration of three Growatt MIN10000TL-X inverters into the farm's power infrastructure. System has 78 modules, model SS-560-12MDH, manufactured by SUNOVA, each inverters have 28 modules totally 43.68 kWp (**Figure 4**).

a. Successfully configured and integrated three Growatt MIN10000TL-X inverters into a solar power system, collaborating with cross-functional teams to achieve a total capacity of 43.68 kWp.

b. Led the planning and execution of the solar panel array, including the installation and wiring of 78 SUNOVA SS-560-12MDH solar modules, optimizing their placement for maximum sunlight exposure and energy efficiency.

c. Conducted detailed site assessments to identify optimal locations for solar panel arrays and inverter placement, taking into account factors such as sunlight exposure, shading, and available space.

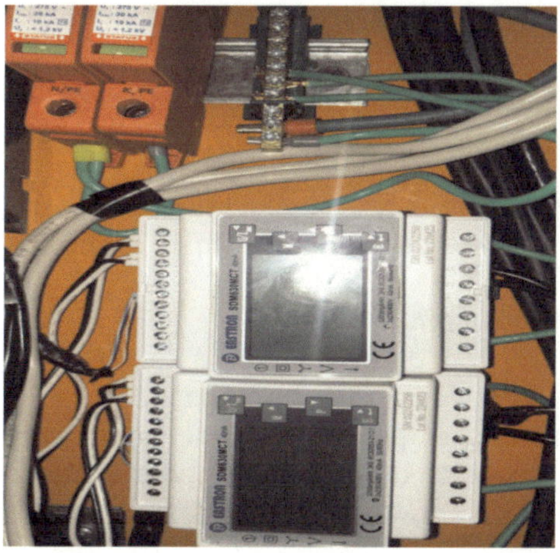

Figure 4.
Smart meter by Growatt applied to the plant. Source: Author himself.

d. Designed and implemented the DC string configuration for the solar modules, ensuring compliance with inverter specifications and maximizing the system's overall performance.

e. Ensured adherence to local electrical codes and safety regulations throughout the entire system configuration process, prioritizing safety and efficiency.

f. Configured Growatt MIN10000TL-X inverters based on manufacturer guidelines, including settings for power output, grid synchronization, and communication interfaces.

g. Conducted thorough testing and commissioning procedures to validate the functionality and reliability of the entire solar power system.

7.3.2 Challenges faced

This part addresses the challenges encountered during the installation phase, such as grid instability, environmental considerations, and any regulatory or compliance issues. Solutions and strategies employed to overcome these challenges are discussed. Among various responsibilities, information management's operational tasks encompass ensuring the safety of the grid against external threats. It is essential to address internal disruptions effectively to maintain a secure power supply. This is especially crucial when considering the 40.2 million homes in Germany [12].

7.4 Integration of smart meters for efficient energy management

7.4.1 Importance of smart meters

Smart meters play a crucial role in monitoring, controlling, and optimizing energy consumption. This section explores the integration of smart meters into the grid zero system to enhance energy management capabilities.

a. By helping consumers to understand their energy usage and make informed decisions about how to reduce it, smart meters can help to lower energy bills

b. Grid zero systems can provide greater energy security by reducing reliance on the traditional power grid, which can be vulnerable to outages and disruptions.

c. By promoting the use of renewable energy, grid zero systems can help to reduce greenhouse gas emissions and combat climate change.

7.4.2 Real-time monitoring and control

The case study highlights the features and benefits of real-time monitoring and control enabled by smart meters, providing farmers with valuable insights into their energy usage patterns. Below is a description of the connectivity of smart meters used on the plant. **Table 1** describes protocol of communication on the plant and interface.

Subsequently, elaborate on how it is employed from the consumer's standpoint for predicting renewable energy needs and managing energy. The sophistication of smart grids makes them increasingly vulnerable to cyber threats due to the monitoring and

Communication	
Interface	RS482
Baud rate	9600
Communication protocol	Modbus RTU

Table 1.
Smart meter communication protocol.

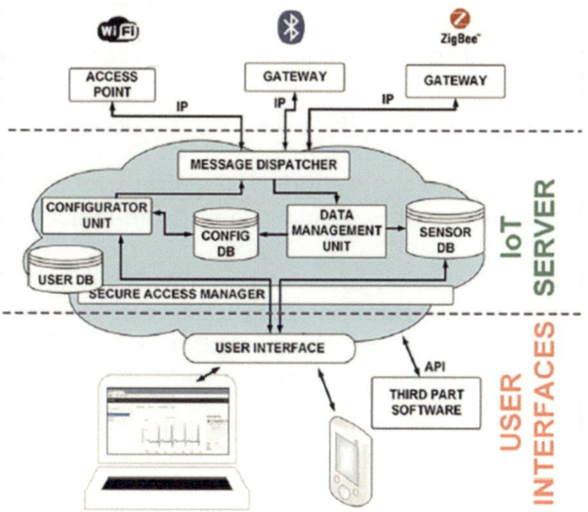

Figure 5.
IoT in smart grid. Source: Ref. [7].

Measurement (Un)	Power export (kW)	Limit inverter (kW)	Percentage (%)
1	80	8	80
2	60	6	60
3	40	4	40

Table 2.
Measurements taken on site.

scrutiny of the Internet of Things (IoT) being conducted through public network connections, as illustrated in **Figure 5**.

Therefore, below demonstrates the characteristic of the smart meter used in the plant to manage the inverter injection power, below, and **Table 2** shows the measurements made "on site."

- Real-time monitoring: Growatt smart meters provide real-time insights into your energy consumption. With instantaneous data updates, users can gain a comprehensive understanding of how electricity is being utilized within their premises.

- Remote accessibility: offering convenience at your fingertips, these smart meters can be accessed remotely. Through a user-friendly interface, you can monitor and analyze energy usage from the comfort of your smartphone or computer.

- Accurate measurement: Growatt's commitment to precision is reflected in the accurate measurement capabilities of their smart meters. Users can trust the data generated by these devices to make informed decisions about energy conservation and optimization.

- Data logging: detailed historical data logging allows users to track energy consumption patterns over time. This comprehensive historical perspective empowers users to make data-driven decisions for enhanced energy efficiency.

- Compatibility: Growatt smart meters are designed to seamlessly integrate with other energy management systems, making them a versatile choice for both residential and commercial applications. Compatibility with solar inverters and other renewable energy sources ensures a holistic approach to energy management.

- User-friendly interface: the intuitive user interface of Growatt smart meters simplifies the monitoring process. Users, regardless of their technical expertise, can navigate through the data effortlessly, gaining valuable insights into their energy consumption.

8. Conclusion

In conclusion, the integration of renewable sources with smart meters offers a promising avenue for enhancing energy efficiency and mitigating the challenges associated with power flow inversion. The advantages of renewable sources, such as sustainability and reduced environmental impact, align with the goals of creating a cleaner and more resilient energy infrastructure. When coupled with the real-time monitoring and control capabilities of smart meters, the synergy between these technologies becomes apparent. Smart meters provide critical insights into energy consumption patterns, enabling more informed decision-making for both consumers and utility companies. However, it is essential to acknowledge the potential disadvantages, including concerns about privacy, security, and the need for substantial initial capital investments. Striking a balance between maximizing the benefits of renewable sources and smart meters while addressing these challenges requires careful planning, robust cybersecurity measures, and a commitment to consumer privacy. As we navigate the complexities of our energy landscape, leveraging the advantages and addressing the disadvantages of these technologies is paramount for a sustainable and efficient energy future.

Author details

Effrain Roney Bernardes
University Federal of Uberlândia, Uberlândia, MG, Brazil

*Address all correspondence to: effrainroney@gmail.com

References

[1] MacQuarie University Sydney 2016 "Evaluation is Assessing Achievements. Available from: http://www.staff.mq.edu.au/teaching/ [Accessed: November 25, 2016]

[2] Fehr M. Independent power projects in sub-Saharan Africa: Investment trends and policy lessons. Energy Policy. **108**: 390-424. Available from: http://ufdcimages.uflib.ufl.edu/AA/00/00/03/83/00121/AA00000383_00121_00078.pdf [Accessed: November 16, 2016]

[3] Matos LM. Collaborative smart grids—A survey on trends. Renewable and sustainable Energy Reviews, Energy Policy. **65**:283-294. Available from: http://ufdcimages.uflib.ufl.edu/AA/00/00/03/83/00121/AA00000383_00121_00078.pdf [Accessed: November 16, 2016]

[4] Energieinformationsnetze und -systeme: Bestandsaufnahme und Entwicklungstendenzen. Ein Positions papier der Informationstechnischen Gesellschaft im VDE (ITG)Watson R, Boudreau M, Chen A. Information systems and environmentally sustainable development: Energy informatics and new directions for the IS community. Management Information Systems Quarterly. 2010;**34**:23-38

[5] Atkins P. Smart Meters and Smart-Enabled Prepaid Meters Project. Department of Economic Development and Tourism, Western Cape Government; 2014. Available from: http://ufdcimages.uflib.ufl.edu/AA/00/00/03/83/00121/AA00000383_00121_00078.pdf [Accessed: November 16, 2016]

[6] Araújo E. Canal Solar. 2024. Available from: https://canalsolar.com.br/reprovacao_de_projetos_a_inversao_de_fluxo_de_potencia_e_de_fato_um_problema [Accessed: January 06, 2024]

[7] Vinoth Kumar P, VenkataRavindra K, Kumar TNVLN. Smart grid issues and challenges. International Journal of Recent Technology and Engineering (IJRTE). Sep 2019;**8**(3):4628-4632. DOI: 10.35940/ijrte.C6836.098319. ISSN: 2277-3878

[8] Fehr M. Quality as a fraction of what is considered perfect (in Portuguese), A qualidade como fração do que é perfeito. Informativo CFQ – Órgão Oficial do CFQ, Rio de Janeiro BR. 1992, 1992;**21** (5/6):4 (part 1); **21**(7/8);2 (part 2). Available from: http://manfredfehr.com.br/qualidade1.pdf http://manfredfehr.com.br/qualidade2.pdf [Accessed: December 06, 2016]

[9] Bella G et al. The relationship among CO_2 emissions, electricity power consumption and GDP in OECD countries. Journal of Policy Modeling. 2014;**36**(6):970-985. DOI: 10.1016/j.jpolmod.2014.08.006

[10] Jagstaidt U. Informationsmanagement: Planung, Überwachungund Steuerung der Informationsinfrastruktur. München Oldenbourg Wissenschaftsverlag

[11] Smadi AA, Ajao BT, Johnson BK, Lei H, Chakhchoukh Y, Al-Haija QA. A comprehensive survey on cyber-physical smart grid testbed architectures: Requirements and challenges. Electronics. 2021;**10**:1043

[12] Budka KC, Deshpande JG, Thotton M. Communication Networks for Smart Grids. Vol. 34(4). Springer Verlag; 2014. pp. 347-372. DOI: 10.5840/teachphil201134447 [Accessed: January 14, 2024]

[13] Measurement, control, and monitoring in smart grids using NBIoT. In: Proc. of Sixth IEEE International

Conference on Inventive Computation Technologies (ICICT). pp. 1217-1222 [Accessed: January 14, 2024]

[14] Security and privacy in smart grids: Challenges, current solutions and future opportunities. Energy Policy. **65**:283-294. Available from: https://www.researchgate.net/publication/339815575_Security_and_Privacy_in_Smart_Grids_Challenges_Current_Solutions_and_Future_Opportunities [Accessed: January 14, 2024]

[15] Torriti J, Hassan MG, Leach M. Demand response experience in Europe: Policies, programmes and implementation. Energy. 2010;**35**:1575-1583

[16] Otuoze AO, Mustafa MW, Larik RM. Smart grids security challenges: Classification by sources of threats. Journal of Electrical Systems and Information Technology. 2018;**5**(3): 468-483

[17] Pandey RK, Misra M. Cyber security threats—Smart grid infrastructure. In: 2016 National Power Systems Conference (NPSC); Bhubaneswar. 2016. pp. 1-6

[18] Jahankhani H, Carlile A, Akhgar B, Taal A, Hessami AG, Hosseinian-Far A. Global security, safety and sustainability: Tomorrow's challenges of cyber security. In: 10th International Conference, ICGS3 2015; September 15-17, 2015. London, UK: Springer; 2015

Section 2

Analysis and Modeling in Smart Grids

Chapter 5

Analytical Analysis of Power Network Stability: Necessary and Sufficient Conditions

Javad Soleimani, Reza Farhangi, Gunes Karabulut Kurt and Fatemeh Mechershavi

Abstract

The investigation of the synchronization of Kuramoto oscillators is a crucial applied model for studying harmonization in oscillating phenomena across physical, biological, and engineering networks. This chapter builds on previous studies by exploring the synchronization of Kuramoto oscillators while also conforming to more realistic models. Using the LaSalle Invariance Principle and contraction property, we introduce the necessary and sufficient conditions for frequency synchronization and phase cohesiveness. The novelty of this chapter's contents lies in three key areas: First, we consider a heterogeneous second-order model with non-uniformity in coupling topology. Second, we apply a non-zero and non-uniform phase shift in coupling function. Third, we introduce a new Lyapunov-based stability analysis technique. Our findings demonstrate that heterogeneity in the network and the phase shift in the coupling function are key factors in network synchronization. We present the synchronization conditions based on network graph-theoretical characteristics and the oscillators' parameters. Analysis of the results reveals that an increase in the phase shift and heterogeneity of oscillators will complicate the synchronization conditions. Numerical simulations confirm the validity of our theoretical results. One of the main applications of this study is the development of stability conditions for smart grids with Lossy-Power Network.

Keywords: Power network, smart grid, transmission line, transient stability, kuramoto oscillator, LaSalle's invariance principle, contraction Property, synchronization, lyapunov function

1. Introduction

Synchronization in interconnected networks of heterogeneous or homogeneous oscillators is a pervasive phenomenon in various fields such as biology, physics, chemistry, engineering, and social networks. The electrical network, as a prime example, is one of the most important and complex oscillatory networks constructed by engineering science. In these networks, two main factors, namely oscillator dynamics and network communication graph, lead to different dynamic behaviors.

Therefore, the overall network response in different regimes can be controlled by the following two approaches:

- Designing and tuning an appropriate control system on the controllable parameters of the oscillatory agents.
- Adjusting the network communication graph based on the parameters of the oscillatory agents.

The future of the electrical network is moving towards increased use of renewable energy sources in the form of distributed generation. On the other hand, the penetration of renewable energy sources leads to an increase in imbalances between electricity production and consumption, along with increased complexity in the dynamic behavior and greater heterogeneity of network factors. Therefore, providing a suitable mathematical model for the studied network [1], understanding the network stability status, and presenting control theories aimed at increasing its stability and robustness against various regimes are among the main concerns of control engineers for the network.

Employing a suitable mathematical model for the studied network is a prerequisite for implementing each of the above approaches [1]. One of the most widely used models in oscillatory networks is the model proposed by Kuramoto. This model describes the dynamic behavior of oscillators whose interactions between agents are proportional to the sine of the phase difference between them.

The electrical network is one of the most important and complex oscillatory networks constructed by engineering science. Due to the oscillatory nature of the dynamic equations of this network, there is a close relationship between its mathematical model and the non-uniform Kuramoto model. Based on this, and considering the limitations of classical methods in studying the stability of electrical networks based on analytical theories, in this thesis, by mapping the oscillation equations of the electrical network to the non-uniform Kuramoto model and presenting analytical theories, a stability condition in the form of an explicit and precise relationship based on the parameters and topology of the network graph is obtained. Furthermore, based on the results obtained from the proposed theories, suggestions for improving network stability are presented.

1.1 Notations

Consider the vector $x = (x_1, \dots, x_n)^T$. The values $_{\min}$ and $x_{\max}$ are respectively the smallest and largest elements of this vector. The 2-norm of vector x is represented by $\|x\|_2$, and $\text{diag}(x) \in \mathbb{R}^{n\times n}$ represents the associated diagonal matrix. The notations 0_n and 1_n are defined as column vectors with zeros and ones in all entries, respectively. The sign $\circ$ represents the entrywise product (*Hadamard Product*) of the two vectors, and for a complex number S, the terms $\Re(S)$ and $\Im(S)$ are the real and imaginary parts of S.

A triple set $G = (v, \varepsilon, A)$ is a weighted directed graph where $A \in \mathbb{R}^{n\times n}$ is the *adjacency* matrix, $v = \{1, \dots, n\}$ is the set of vertices, and $\varepsilon \subset v \times v$ is the set of edges. In the *adjacency* matrix, $a_{ij} > 0$ if there is a directed edge from vertex i to j, otherwise $a_{ij} = 0$. The *Laplacian* matrix of G, denoted by L, is defined as $L := \text{diag}\left(\sum_{j=1}^{n} a_{ij}\right) - A$. For the *incidence* matrix $B = H^T \in \mathbb{R}^{n\times|\varepsilon|}$, the entry $B_{kl} = -1$ if edge l is directed to vertex K, and $B_{kl} = 1$ if edge l is directed from vertex K, otherwise $B_{kl} = 0$. Notation $\ker(H)$ is relevant to the *null space* of matrix H. The *complete* graph is defined as a graph where all vertices are connected. If n is equal to the number of vertices of the *complete* graph, the number of edges is given by $n(n-1)/2$. If G is a *connected* graph, then $\ker(H) = \ker(L) = \text{span}(\mathbf{1}_n)$, all non-zero eigenvalues of the *Laplacian* matrix are positive, and $\lambda_2(L)$, the second-smallest eigenvalue of G, is named the *algebraic connectivity*. $\lambda_{\min}(L)$ and $\lambda_{\max}(L)$ correspond to the minimum and maximum eigenvalues of matrix L, respectively. Moreover, $\lambda_2(L) = n$ in a complete and uniformly weighted graph ($a_{ij} = 1$ for all $i \neq j$).

The *torus* $\mathbb{T}^1 = (-\pi, \pi]$ is a set, and angle θ is a point of the *torus* where $\theta \in \mathbb{T}^1$, and the *arc* corresponds to the subset of $\mathbb{T}^1$. The distance $|\theta_i - \theta_j|$ is defined as the minimum length between two angles $\theta_i \in \mathbb{T}^1$ and $\theta_j \in \mathbb{T}^1$. For $\rho \in [0, \pi/2]$, the $\Delta(\rho) \subset \mathbb{T}^1$ is the set of surrounded angles $(\theta_1, \dots, \theta_n) \in \mathbb{T}^n$, where the *arc* with length ρ comprises all the angles $\theta_1, \dots, \theta_n$. Accordingly, for every angle of the set $\Delta(\rho)$, the inequality $\max_{\{i,j\}\in\varepsilon}|\theta_i - \theta_j| < \rho$ is satisfied. For $\in \mathbb{R}^n$, we define the vector-valued function $\mathbf{sin}(x) = (\sin(x_1), \dots, \sin(x_n))$ and the sinc function sinc : $\mathbb{R} \to \mathbb{R}$ by $\text{sinc}(x) = \sin(x)/x$.

1.2 Classical methods for stability analysis

The classical methods used to determine the stability of a network in the early years of its development were based on available computational tools, such as the ability to maintain network stability during numerical simulations of worst-case scenarios in the design or based on the direct Lyapunov method [2–6]. However, gradually, with the application of these methods in real networks, some inefficiencies in the performance of classical methods became apparent such as the following items [7, 8]:

1. The need for complex and time-consuming numerical computations in numerical simulations.

2. Inefficiency of the direct Lyapunov method in networks with transmission losses. Considering the energy function as the equation $E = \sum_{i=1}^{n}\left(M_i\dot{\theta}_i^2/2\right) - \sum_{\{i,j\}\in\varepsilon}^{n} P_{ij}\cos\theta_{ij} - \sum_{i=1}^{n}\omega_i\theta_i$ (where $\theta = (\theta_1, \dots, \theta_n)^T$ and $\theta_{ij} = \theta_i - \theta_j$), and the oscillation equations for lossless electrical networks as $M_i\ddot{\theta}_i + D_i\dot{\theta}_i = \omega_i - \sum_{j=1}^{n} P_{ij}\sin(\theta_i - \theta_j) \quad \forall \;\; i \in \{1, \dots, n\}$, the derivative of the energy function is obtained as $\dot{E} = -\sum_{k=1}^{n} D_i\dot{\theta}_i^2 \leq 0$. In this case, the network is locally stable [9]. This function can be extended to networks with transmission losses as $E_{loss} = \sum_{i=1}^{n}\left(M_i\dot{\theta}_i^2/2\right) - \sum_{\{i,j\}\in\varepsilon}^{n} P_{ij}\cos\left(\theta_{ij} + \phi_{ij}\right) - \sum_{i=1}^{n}\omega_i\theta_i$, which is obtained for $\dot{E} = -\sum_{i=1}^{n} D_i\dot{\theta}_i^2 - 2\sum_{\{i,j\}\in\varepsilon}^{n} P_{ij}\dot{\theta}_j\cos\theta_{ij}\cos\phi_{ij}$. As observed, this function is not always decreasing and cannot guarantee network stability.

3. The inability to employ network controllers and topological information for feedback control transmission in stability studies, and the excessive conservatism in these studies.

1.3 Transient stability study through Kuramoto model

In recent years, the increasing random disturbances in the network due to the growing demand for electricity consumption and the penetration of renewable energy sources have highlighted the limitations of classical methods for analyzing network stability. Alongside the consequences of instability in electrical networks and the need to describe network stability through explicit mathematical relationships based on network parameters, the main motivation for using structure-oriented analytical methods as alternatives to classical analytical methods has been the development of measurement and computational technologies that provide a suitable platform for implementing their results in the network [10]. In this regard, a review of the history of studies shows that the oscillatory nature of electrical networks and the similarity between the model of these networks and the Kuramoto oscillatory model have led to a strong tendency to use this model for analyzing the synchrony of electrical networks.

In [11] Jadbabaie demonstrated that the region $\theta \in \Delta(\pi/2)$ for $P > 0$ is an invariant set for a first-order oscillatory network with a complete graph and equal natural frequencies, $\dot{\theta}_i = \omega_i + (P/n)\sum_{j=1}^{n} \sin(\theta_j - \theta_i)$, state trajectories of each agents simultaneously converge at rate $(2P/\pi n)\lambda_2(L)$. Additionally, for a network with agents having asynchronous natural frequencies and a complete graph, the phase coherence condition of agents is obtained as $P \geq P_L: = 2\sqrt{n}\|\omega\|_2/\lambda_2(L)$ and $P \geq P_L: = n\pi^2\lambda_{\max}(L)\|\omega\|_2/4\lambda_2(L)^2$, respectively, in $\theta \in \Delta(\pi/2)$. Further studies by Chopra and Spong obtained the necessary condition for synchrony in terms of $P \geq P_{ess} = n(\omega_{\max} - \omega_{\min})/E_{\max}$ (with $E_{\max}$ being a computational parameter of the network model) by studying the dynamics of rotational frequency difference of oscillators [12]. This condition becomes the sufficient condition for synchrony under extreme conditions $P_{ess} = (\omega_{\max} - \omega_{\min})/2$. Furthermore, in this reference, the phase coherence condition of agents in $\theta \in \Delta(\rho)\forall\rho \in \pi/2$ is obtained as $P \geq P_{suf} = n(\omega_{\max} - \omega_{\min})/\cos(\pi/2 - \rho)$. To complement the research [12], Choi et al. used Dini derivatives to derive necessary conditions for stability in phase-locked states, taking into account the initial conditions of the model in the form of $P \geq (\omega_{\max} - \omega_{\min})/\sin(\theta_{\max} - \theta_{\min})$. In [13], they presented these conditions as a function of the initial phase of the model. In [13] sufficient conditions for frequency synchronization were also obtained. In [14], Ha et al. built upon the results from [13] and investigated the effects of dynamic frequency natural factors and phase shifts on the synchronization of oscillators in a limited set of oscillators. They derived stability conditions as functions of the initial phase of the oscillators, the amount of phase shift, and the intensity of network communications. It should be noted that the impact of losses in the dynamic model of transmission lines appears as phase delay. Previously, studies on the stability of oscillatory networks with phase shifts have been evaluated in [15–21]. In [15], by introducing the parameter link frustration and minimizing the values of dynamic states, the dynamic behavior of networks with different topologies was investigated. In [16], simulations showed how changing the value of phase shift can affect agent asynchrony. However, it is important to note that the results obtained from these studies rely on numerical approaches.

In [21, 22], as one of the first steps in utilizing oscillatory models in the stability analysis of electrical networks, the threshold level of disturbances for each node that leads to loss of synchrony was calculated through numerical analyses, and the relationship between this threshold level and the topological characteristics of the network graph and the time of loss of synchrony was determined. The results obtained from these analyses indicate a clear relationship between the dynamic and topological parameters of the network, such that an increase in the degree of graph nodes increases the minimum resolvable disturbances in the network. These results can be used in the design or upgrading of networks with the aim of improving network stability. Additionally, Dörfler and his colleagues in [22–32], analyzed the stability of electrical networks using oscillatory models. Building upon the results of studies [33–36], Dörfler and Bullo in [26, 27] derived sufficient conditions for transient stability in electrical networks based on the topological characteristics of the network graph, equipment parameter values, and initial conditions, by considering the assumption of a small ratio of inertia parameter to damping coefficient of network components, $M_{\max}/D_{\min}$, and employing techniques based on Singular Perturbation and Lyapunov stability.

In [28], Dörfler and his colleagues presented a more explicit condition for achieving phase coherence among agents by analyzing the first-order oscillatory model on synchronous manifolds. Based on the results obtained for the first-order model, it was shown that by satisfying the condition $\left\|B^T L^\dagger \omega\right\|_\infty \leq \sin(\rho)$, a non-trivial region for the phase difference of the agents $|\theta_i - \theta_j| \leq \rho \leq \pi/2$ can be established.

Given the importance of frequency control and synchronization in microgrids, the adjustment of network control systems is of great importance. In [25, 26, 28, 31, 32, 37–43], stability studies of microgrids have been conducted using oscillatory models. In [29] and [30], the frequency stability of an inverter-based microgrid with Power-Frequency Droop Controllers is investigated assuming $M_i = 0$, and the necessary and sufficient conditions for synchronization in the microgrid are obtained. In [44], the impact of network decentralization on stability is studied numerically in a power network with distributed energy resources. In [45, 46], using model [47], the necessary and sufficient condition for transient stability in an electric network with controllable injection power to oscillators is presented as $\dot{\omega}_i = K_i \sum_{j=1}^{n} b_{ij} \left(\dot{\theta}_j - \dot{\theta}_i\right)$. In [38], by employing theories from [27, 30, 48–50], and designing a distributed adaptive droop controller to enhance the algebraic connectivity of the network, the necessary and sufficient condition for stability of low-inertia microgrids is developed. In [51], by using the Kron reduction process and adjusting parameters including generator droop and network damping coefficient, the eigenvalues of the system are controlled for network stability. In [39], a distributed proportional-integral controller is employed to analyze the synchronization and phase tracking problem.

In [52], using Gronwall's inequality, the convergence rate is expressed as a function of inertia, coupling strength, and natural frequency for phase-shift-free communications. In [53], the same studies are conducted for a second-order model with phase shift in the communication function. However, the results of these studies are valid for networks with homogeneous agents, uniform communications, and equal phase shifts. Meanwhile, Choi et al. calculate the synchronization conditions of this model in [54] using energy functions as a function of the initial topology, distribution of natural frequencies, and agent inertia for networks with homogeneous agents. In [55], the same studies are conducted for networks with heterogeneous oscillatory agents under certain assumptions. Thanh Long et al. extend their previous stability analysis methods in [56–59] and classical stability analysis methods in [60–65] for electric networks with transmission line losses in [8]. They develop a Lyapunov function for networks with lossy transmission lines using an iterative approach based on initial conditions. Bo Li et al. in [66] present an optimization algorithm for the coupling strength and input power to agents in order to increase the value of $\lambda_2(L(\theta^*))$ in a linear model of a power network. In [67], Grzybowsk et al. provide an estimation of the synchronization basin region in a power network based on the second-order Kuramoto model, considering assumptions such as equal damping of oscillatory agents and neglecting transmission line resistances.

In [68, 69], an estimation of the stable synchronization region of an electric network is provided using the second-order Kuramoto model, based on the Lojasiewicz and Gronwall's inequalities and the assumption $m_i/d_i = m_j/d_j$ for $i \neq j$. In [70], the same studies are conducted without assumptions and limitations.

The authors in [71] continue their studies from [10, 27, 32, 72], and their previous work in [40, 41], by introducing a Lyapunov function and using the second-order Kuramoto model to obtain a sufficient condition for synchronization in an electric network with lossless transmission lines. In [72], local stability analysis for the electric network model introduced in [47] is also conducted. In [73], Yufeng et al. calculate the critical outage time and return the network to a stable state using the estimated synchronization region for network frequency synchronization in reference to [68] through numerical analysis. In [74–79], analysis of multi-zone or cluster oscillatory networks is conducted.

Based on the above, the study of stability in oscillatory network has shown that by performing the following transformations in the used model, the results of these studies have become closer to reality over the past two decades:

- First-order models to second-order models;
- Homogeneous agents to heterogeneous agents;
- Uniform coupling graph to non-uniform coupling graph;
- Phase-shift-free coupling to the coupling topology with phase shift.

1.4 Objectives and scope of the chapter

As mentioned earlier, it is not possible to determine the stability condition of a network explicitly based on the characteristics of the agents and the Laplacian matrix of the network using classical stability analysis methods. Therefore, one of the open issues in power system conferences and forums is to determine explicit and accurate conditions based on the network characteristics that guarantee transient stability.

Reviewing the conducted activities shows that the developed theories in recent studies have usually involved assumptions about the oscillatory network agents or specific conditions in the network graph. As a result, the results of each reference have been developed based on previous studies, and the assumptions have gradually decreased. Generally, these assumptions can be categorized as follows:

- Considering homogeneous parameters of network agents such as damping coefficient and inertia in [13, 14, 53].
- Assuming uniform coupling strengths between agents in [11–14, 53].
- Assuming overdamping conditions in the network and transforming the oscillation equations from second-order models to first-order models using singularity perturbation techniques in [26–28, 30, 45, 46].

- Local stability studies based on linearized network models in [51, 72, 79, 80].
- Assuming equal inertia-to-damping ratio for network agents in [55, 68, 69].
- Developing results based on analyzing the numerical simulation results in [16, 22, 44, 66, 73, 74].
- Considering small inertia of oscillatory network agents in [25–28, 31, 32, 37–46] stability analysis of microgrids that cannot be applied to electric networks with conventional power plants.
- Analyzing lossless transmission line network models except for references [8, 14, 16, 17, 27, 52, 53].
- Analyzing network stability without considering the effect of control signals in the network except for references [25, 29, 30, 39, 45–46, 51].

In more comprehensive models, such as [8, 52, 53, 67, 68–71, 73], assumptions and constraints have been introduced. Some examples include:

- Calculation of the Lyapunov function based on iterative algorithms [8].
- Studies on lossless networks [52].
- Assuming equal inertia-to-damping ratio for network agents in [55].
- Limited number of oscillators in [68, 69].
- Employing the assumption of a complete graph in a network with an incomplete graph [71].

Therefore, it can be claimed that many of the theories presented in recent references have provided valuable results in complementing previous studies. In the following sections, the necessary and sufficient conditions for the stability of an electrical network with significant losses, heterogeneous oscillatory agents, and non-uniform communications are presented based on explicit and clear relationships between the characteristics of the network graph, initial conditions, and properties of oscillatory agents.

2. Fundamentals of power network transmission

2.1 Energy balance in generators

In the electric network oscillating model, the dynamics of the network's generators are usually described using state variables such as phase angle δ_i and rotor angular velocity $d\delta_i/dt$. The value of Ω corresponds to the nominal frequency of the network and is equal to $\Omega = 2\pi \times 50\ Hz$ Hz or $\Omega = 2\pi \times 60\ Hz$ Hz. Therefore, the phase angle of the generators will be:

$$\delta_i(t) = \Omega t + \Theta_i(t) \tag{1}$$

Where Θ_i represents the phase difference between the generator and the reference generator Ωt. During the rotation of the generators in the electric network, the amount of dissipated power is obtained by the expression $P_{diss,i} = K_{D,i}\dot{\delta}_i^2$, where $K_{D,i}$ is equal to the overall losses coefficient, friction coefficient or other power loss factors. The amount of kinetic energy and mechanical power in the electric generators are equal to $E_{kin,i} = I_i\dot{\delta}_i^2/2$ and $P_{acc,i} = dE_{kin,i}/dt$ respectively, where I_i is equivalent to the moment of inertia of the i-th generator. If there is power flow between two nodes i and j, then the transfer power between node i and j is denoted by $P_{trans,ij}$. In order to achieve power balance in a generator, it is necessary for the injected power $P_{source,i}$ to the generator to be equal to the sum of kinetic power, dissipated power, and transferred power from the generator, $P_{trans,i} = \sum_{j=1}^{N} P_{trans,ij}$. In other words, $P_{source,i} = P_{diss,i} + P_{acc,i} + \sum_{j=1}^{N} P_{trans,ij}$. Substituting the power balance equation in terms of generator states, we obtain:

$$P_{source,i} = K_{D,i}\dot{\delta}_i^2 + I_i\ddot{\Theta}_i\dot{\delta}_i + P_{trans,i} \tag{2}$$

Utilizing $\delta_i(t) = \Omega t + \Theta_i(t)$ in the prior equation results:

$$\begin{aligned} P_{source,\,i} &= K_{D,i}\left(\Omega + \dot{\Theta}_i\right)^2 + I_i\ddot{\Theta}_i\left(\Omega + \dot{\Theta}_i\right) + P_{trans,\,i} \\ &= K_{D,i}\Omega^2 + 2K_{D,i}\Omega\dot{\Theta}_i + K_{D,i}\dot{\Theta}_i^2 + I_i\Omega\ddot{\Theta}_i + I_i\ddot{\Theta}_i\dot{\Theta}_i + P_{trans,\,i} \end{aligned} \tag{3}$$

Where, $2K_{D,i}\Omega$ is equal to the damping coefficient. Assuming that the impact of disturbances on the rotational frequency of network generators compared to the nominal frequency of the network $\dot{\Theta}_i \ll \Omega$, then the terms $I_i\ddot{\Theta}_i\dot{\Theta}_i$ and $K_{D,i}\dot{\Theta}_i^2$ can be neglected compared to $I_i\Omega\ddot{\Theta}_i$ and $K_{D,i}\Omega^2$ [81]. The power balance equation in network generators will be as follows:

$$I_i\Omega\ddot{\Theta}_i = P_{source,i} - K_{D,i}\Omega^2 - 2K_{D,i}\Omega\dot{\theta}_i - P_{trans,i} \tag{4}$$

To simplify, we define $M_i: = I_i\Omega$ as the moment of inertia, $P_{m,i}: = P_{source,i} - K_{D,i}\Omega^2$ as the net power provided to the i-th generator, and $D_i: = 2K_{D,i}\Omega$ is defined as the damping coefficient for the i-th generator. Thus the power balance network generators can be simplified to the following:

$$M_i\ddot{\Theta}_i = P_{m,i} - D_i\dot{\Theta}_i - P_{trans,i} \tag{5}$$

2.2 Electric network model reduction

To calculate $P_{trans,i}$ as Eq. (5), the real part of matrix $\Re(V \circ I^*)$, the vectors of node voltages V, and the injected current to the nodes I are necessary. The value of the vector I is obtained the reference node form as $I = Y_{net}V$, where Y_{net} is the admittance matrix of the network. With the assumption of static loads in the electric network, the total of N network nodes can be divided into dynamic nodes N_G and static nodes N_G. So the network admittance matrix Y_{net} can be divided into four separate sections as follows:

$$Y_{net} = \begin{bmatrix} Y_{GG} & Y_{LG} \\ Y_{LG} & Y_{LL} \end{bmatrix} \tag{6}$$

The injected current values to each of the network nodes will be as follows:

$$\begin{bmatrix} I_G \\ I_L \end{bmatrix} = Y_{net} \begin{bmatrix} V_G \\ V_L \end{bmatrix} \rightarrow \begin{matrix} I_G = Y_{GG}V_G + Y_{LG}V_L \\ I_L = Y_{LG}V_G + Y_{LL}V_L \end{matrix} \tag{7}$$

Where $V_G = [V_1 \ \ ... \ \ V_{N_G}]^T$ represents the node voltages of the network generators, $V_L = [V_{N_G+1} \ \ ... \ \ V_N]^T$ is the node voltages of the network loads, $I_G = [I_1 \ \ ... \ \ I_{N_G}]^T$ is the injected currents to the network generators, and $I_L = [I_{N_G+1} \ \ ... \ \ I_N]^T$ represents the injected currents to the network loads. In the following, the assumption of zero resistance for transmission lines has been disregarded, and an attempt has been made to use a more realistic model of the network in stability analyses.

As mentioned before, assuming the static nature of loads compared to the dynamics of network generators, the network loads can be modeled as admittances shunted to the reference node in an N_L node model. It is worth noting that since transient stability studies are conducted in this chapter and considering the time frame for transient stability studies, the assumption of static loads is reasonable. In this case, by applying the Kron reduction method, we can obtain the voltage V_L in terms of V_G and substitute it into equation I_G.

$$V_L = Y_{LL}^{-1}I_L - Y_{LL}^{-1}Y_{LG}V_G \tag{8}$$

$$I_G = Y_{GG}V_G - Y_{LG}\left(Y_{LL}^{-1}Y_{LG}V_G\right) + Y_{LG}Y_{LL}^{-1}I_L = \left(Y_{GG} - Y_{LG}Y_{LL}^{-1}Y_{LG}\right)V_G + Y_{LG}Y_{LL}^{-1}I_L \tag{9}$$

So the reduced admittance matrix of the network will be $Y_{red,g} = Y_{GG} - Y_{LG}Y_{LL}^{-1}Y_{LG}$. By extracting the shunt admittance due to static loads from the matrix $Y_{red,g}$ and inputting it into the generator nodes, Eq. (7) will be $I_G = Y_{red}V_G$. In this regard, the net injected power by the generators into the network will be given by the following:

$$S_{trans} = V_G \circ I_G^* = V_G \circ \left(Y_{red}^* V_G^*\right) = V_G \circ Y_{red}^* V_G^* + S_{L2G} \tag{10}$$

Where $S_{L2G} = V_G \circ \left(Y_{LG}Y_{LL}^{-1}I_L\right)^*$ represents the apparent power effect of network loads on network generator nodes. By substituting the voltage phasors of the generator nodes and the reduced admittance matrix, we have:

$$S_{trans,i} = S_{L2G,i} + \sum_{j=1}^{N_G} |V_i||V_j||Y_{red,ij}|e^{i(\Theta_i - \Theta_j - \varphi_{ij})} \tag{11}$$

Where Θ_i and Θ_j respectively represent the voltage angles of generator i and j, and φ_{ij} represents the phase angle of the reduced admittance matrix between two nodes $i, j \in \{1, \ldots, N_G\}$, which is obtained from the equation $\varphi_{ij} = \text{atan}\left(\Im\left(Y_{red,ij}\right)/\Re\left(Y_{red,ij}\right)\right)$. Taking into account the power effect caused by network loads on network generator nodes as $P_{L2G,i} = \Re(S_{L2G,i})$, the active transfer power from generator i will be as follows:

$$P_{trans,i} = P_{L2G,i} + \sum_{j=1}^{N_G} |V_i||V_j||Y_{red,ij}| \cos\left(\Theta_i - \Theta_j - \varphi_{ij}\right) \tag{12}$$

Employing the trigonometric identity $\sin(x) = \cos(x - \pi/2)$ and defining the parameter $\phi_{ij} = -\varphi_{ij} + \pi/2$, the equation for active transfer power from generator i will be transformed to:

$$P_{trans,i} = P_{L2G,i} + \sum_{j=1}^{N_G} |V_i||V_j||Y_{red,ij}| \sin\left(\Theta_i - \Theta_j + \phi_{ij}\right) \tag{13}$$

Moreover, applying the identities $\sin(\alpha \pm \beta) = \sin(\alpha)\cos(\beta) \pm \cos(\alpha)\sin(\beta)$ and $\cos(\alpha \pm \beta) = \cos(\alpha)\cos(\beta) \mp \sin(\alpha)\sin(\beta))$, the relationship between the angle ϕ_{ij} and the reduced admittance matrix will be as follows:

$$\tan\left(\phi_{ij}\right) = \tan\left(-\varphi_{ij} + \frac{\pi}{2}\right) = \frac{\sin\left(-\varphi_{ij} + \frac{\pi}{2}\right)}{\cos\left(-\varphi_{ij} + \frac{\pi}{2}\right)} = \frac{\sin\left(\frac{\pi}{2}\right)\cos\left(-\varphi_{ij}\right)}{-\sin\left(\frac{\pi}{2}\right)\sin\left(-\varphi_{ij}\right)} = \frac{\Re\left(Y_{red,ij}\right)}{\Im\left(Y_{red,ij}\right)} \tag{14}$$

Lemma 1: Applying the Kron reduction method in an electrical network, the off-diagonal and diagonal elements of the reduced admittance matrix Y_{red} will be located in the regions $(\pi/2, \pi)$ and $(-\pi/2, 0)$ (respectively)

Proof: In an electrical network, the off-diagonal and diagonal elements of the Y_{net} matrix, representing non-diagonal and diagonal elements, are located in the regions $(\pi/2, \pi)$ and $(-\pi/2, 0)$ respectively. Utilizing the Kron reduction method and eliminating the node corresponding to the load with index $p \in \{N_G + 1, \dots, N\}$ in the Y_{net} matrix, the reduced element $Y_{red,jk}$ can be obtained from the following equation:

$$Y_{red,jk} = Y_{net,jk} - \frac{Y_{net,jp} Y_{net,pk}}{Y_{net,pp}} \tag{15}$$

The elements of the network admittance matrix are represented in phasor form as $Y_{net,jk} = |Y_{net,jk}|\measuredangle\Theta_{jk}$, $Y_{net,jp} = |Y_{net,jp}|\measuredangle\Theta_{jp}$, $Y_{net,pk} = |Y_{net,pk}|\measuredangle\Theta_{pk}$, and $Y_{net,pp} = |Y_{net,pp}|\measuredangle\Theta_{pp}$, where their phases lie in the regions $\Theta_{pk}, \Theta_{jp}, \Theta_{jk} \in (\pi/2, \pi)$ and $\Theta_{pp} \in (-\pi/2, 0)$. Based on these angles, the phase of the element jk-th in the reduced matrix Y_{red} can be expressed as a phasor, $Y_{red,ij} = |Y_{red,ij}|\measuredangle\varphi_{red,jk}$ $\forall$ $\varphi_{red,jk} \in (\pi/2, \pi)$. In other words, $\Re\left(Y_{red,ij}\right)$ and $\Im\left(Y_{red,ij}\right)$ will have negative and positive values, respectively. Therefore, the phases of the off-diagonal and diagonal elements of the Y_{red} matrix will lie in the regions $(\pi/2, \pi)$ and $(-\pi/2, 0)$, respectively. This concludes the proof of this Lemma.

As mentioned in Lemma 1, in the reduced network admittance matrix of the electrical network, without considering losses in transmission lines, the angle $\varphi_{ij} = \text{atan}\left(\Im\left(Y_{red,ij}\right)/\Re\left(Y_{red,ij}\right)\right)$ will fall within the range $\varphi_{ij} \in (\pi/2, \pi)$. Consequently, the angle $\phi_{ij} = -\varphi_{ij} + \pi/2$ will also lie within the range $\phi_{ij} \in (0, \pi/2)$. By substituting

$P_{trans,i}$ from Eq. (13) into Eq. (5), the oscillation equations of the electrical network generators can be expressed as follows:

$$M_i\ddot{\Theta}_i = P_{m,i} - P_{L2G,i} - D_i\dot{\Theta}_i - \sum_{j=1}^{N_G} |V_i||V_j||Y_{red,ij}| \sin\left(\Theta_i - \Theta_j + \phi_{ij}\right) \tag{16}$$

It is evident that two nodes, denoted as i and j, within the reduced network or dynamic model, will be linked by an edge if there is a path from node i to node j in the dynamic-algebraic model. Consequently, the reduced graph of a network with a connected graph will exhibit a complete graph structure. Additionally, node i in the reduced network will have an internal loop if there is a path from node i to a node connected to the reference node in the original network, and the reference node is not included as a separate node in the admittance matrix of the network Y_{net}. Hence, when the original graph is connected and one of the removed nodes is connected to the reference node, all nodes in the reduced network will be connected to the reference node.

3. Transient stability analysis

According to the previous section, by reducing the dimension of the dynamic-algebraic model of an electrical network to a dynamic model, the power transfer between nodes of the dynamic network will be a function of the sine of the phase angle difference between node voltages and the phase shift ϕ_{ij}. Based on Eq. (16), the swing equations of the generators in the electrical network can be presented as follows:

$$M_i\ddot{\Theta}_i + D_i\dot{\Theta}_i = \omega_i - \sum_{j=1, j\neq i}^{N_G} P_{ij} \sin\left(\Theta_i - \Theta_j + \phi_{ij}\right) \qquad i \in \{1, \ldots, N_G\} \tag{17}$$

Where, $\omega_i = P_{m,i} - P_{L2G,i}$, $P_{ij} = |Y_{red,ij}||V_i||V_j|$, $\phi_{ij} = -\varphi_{ij} + \pi/2$, $\quad \phi_{ij} \in (0, \pi/2)$, and $\varphi_{ij} = \text{atan}(\Im(Y_{red,ij})/\Re(Y_{red,ij}))$, Y_{red} is the reduced admittance matrix of the network, Θ_i is the phase angle of voltage V_i, $P_{m,i}$ is the mechanical power injected into generator i, $P_{L2G,i}$ is the effect of network loads on generator i, and N_G is the number of generators in the network.

The presented model in Eq. (17) corresponds to the second-order non-uniform Kuramoto model with a non-zero phase shift $\phi_{ij} \neq 0$. Considering this model and the results obtained in the previous section, the transient stability of the electrical network is investigated in this chapter. Based on the obtained results, an explicit and direct method is presented to determine the necessary and sufficient condition for the frequency synchronization and phase coherence of the network generators. Dividing both sides of Eq. (17) by the inertia factors, the dynamic of the oscillatory agent i in the electrical network can be obtained as follows:

$$\ddot{\Theta}_i = \frac{\omega_i}{M_i} - \frac{D_i}{M_i}\dot{\Theta}_i - \sum_{j=1, j\neq i}^{N_G} \frac{P_{ij}}{M_i} \sin\left(\Theta_i - \Theta_j + \phi_{ij}\right) \tag{18}$$

It is evident that the variation in the inertia values of network components leads to the asymmetry of coupling strengths in the dynamic model of oscillating agents.

According to Eq. (18), for two oscillating agents i and j that are coupled, the coupling strength in the dynamics of agent i is proportional to P_{ij}/M_i while the coupling strength in the dynamics of agent j is proportional to P_{ji}/M_j. This complexity arises from the analytical computations performed on the dynamic model of oscillating agent, considering the variation of inertia values. However, in order to analyze the dynamic behavior of the network model and determine sufficient conditions for phase coherence, it is necessary to specify the dynamic model of the frequency differences of oscillating agents in the electrical network. Through employing Eq. (18), the dynamic model of the network can be derived as follows:

$$\frac{d}{dt}(H\dot{\Theta}) = HM^{-1}\omega - HM^{-1}D\dot{\Theta} - HM^{-1}H^T\text{diag}(P_{ij})\,\mathbf{sin}\,(H\Theta + \phi) \tag{19}$$

where $\phi = \text{diag}\left(\phi_{ij}\right)\mathbf{1}_{|\varepsilon|} \in \mathbb{R}^{|\varepsilon|\times 1}$ is equivalent to a column vector. As mentioned, except for [27], the cases where the necessary and sufficient condition for frequency synchronization and phase coherence of the first or second-order Kuramoto model is explicitly presented in the form of analytical methods, are related to conditions where symmetry is established in agent communications, and references [12, 33, 52–54] are among the most important of these references. The theorem presented in [27] also provides a sufficient condition for phase coherence of agents in an electrical network with an analytical approach. However, the results of this reference are reliable only under the condition that the value of M_i/D_i for small network generators is considered.

In this regard, and with the aim of providing a comprehensive theory for phase coherence of agents in an electrical network, first, the necessary condition for frequency synchronization of the network based on physical and topological characteristics is presented in the form of theorem 1. Then, using two approaches, the direct Lyapunov method and the compressibility property, a sufficient condition for boundedness of the phase difference between generators is extracted in sections 3–2 and 3–3, respectively, in the form of theorems 2 and 3.

3.1 Necessary condition for synchronization

In continuation of studies [11–13, 26, 33],this section provides the "necessary condition" for synchronization in the electrical network with significant losses in transmission lines and an asymmetric communication matrix.

Theorem 1. Consider the power network oscillators' phase difference dynamic model Eq. (17), where the coupling strength $\mathrm{P} = \mathrm{P}^{\mathrm{T}}$ and $B = H^T \in \mathbb{R}^{N_G\times|\varepsilon|}$ is the incidence matrix of the complete graph. The symmetric phase shift ϕ_{ij} and the phase Θ_i are located in the regions $\phi_{ij} \in (0, \pi/2)$ *and* $\{\Theta(t) \in \Delta(\rho) : \rho \in (0, \pi/2)\}$. In the synchronization conditions, the network states are located on the synchronous manifold $A = \{(\Theta, f)\ \ \forall\ \ \theta \in \Delta(\rho), f = \omega_{syn}\}$, and the dynamic model will be simplified to the following algebraic equation:

$$D_i\omega_{syn} = \omega_i - \sum_{j=1, j\neq i}^{N_G} P_{ij}\sin\left(\Theta_i - \Theta_j + \phi_{ij}\right) \qquad i \in \{1, \ldots, N_G\} \tag{20}$$

By defining the vector X with element i as $X_i = \sum_{j=1, j\neq i}^{N_G} P_{ij}\sin\left(\phi_{ij}\right)\cos(\Theta_i - \Theta_j)$ and parameters $S_\omega = \sum_{i=1}^{N_G} X_i$ *and* $\phi_{\max} = \max\left\{\phi_{ij}\right\}$, the following propositions will hold:

Statement 1) The necessary condition for synchronization in an electrical network, assuming that the phase of all agents is limited to the set $\{\Theta(t) \in \Delta(\rho) : \rho \in (0, \pi/2)\}$, will be as follows:

$$
\begin{aligned}
(\Upsilon_{\max} - \Upsilon_{\min}) \le K_{suf}\left(\deg_i + \deg_j\right) \quad for \quad all\{ij\} \in \varepsilon \ \Upsilon_{\max} \\
= \max\left\{\frac{\omega_i}{D_i}\right\}, \\
\Upsilon_{\min} = \min\left\{\frac{\omega_i}{D_i}\right\},\ K_{suf} = \sin(\rho + \phi_{\max})/D_{\min} \quad for(\rho + \phi_{\max}) \\
\le \left(\frac{\pi}{2}\right) K_{suf} = 1/D_{\min} \quad for(\rho + \phi_{\max}) \ge \left(\frac{\pi}{2}\right)
\end{aligned}
\tag{21}
$$

Statement 2) The actual frequency of oscillations ω_{real}, is determined by Eq. (22), where ω corresponds to the nominal frequency of the network.

$$
\omega_{real} = \omega + \omega_{syn}\ ,\omega_{syn} = \sum_{i=1}^{N_G} \omega_i - S_\omega / \sum_{i=1}^{N_G} D_i \tag{22}
$$

Proof of Statement 1: Synchronization in Eq. (20) requires that the maximum power transfer between nodes through inter-agent communications is greater than the maximum difference in input power to the agents. Substituting the maximum value of the term $\sin\left(\Theta_i - \Theta_j + \phi_{ij}\right)$ and $(\Upsilon_{\max} - \Upsilon_{\min})$ on the right side of the this model, we will reach to the minimum connectivity strength to satisfy the synchronization condition. Considering the assumption of Theorem 1 regarding the phase of the agents $(\Theta_1, \ldots, \Theta_{N_G}) \in \mathbb{T}^{N_G}$ belonging to the region $\Delta(\rho) \subset \mathbb{T}^1\ \forall\ \rho \in [0, \pi/2]$, it is possible to conservatively replace the term $\sin\left(\Theta_i - \Theta_j + \phi_{ij}\right)$ with $\sin(\rho + \phi_{\max})$ or $\sin(\pi/2)$. Therefore, by satisfying the inequality $(\Upsilon_{\max} - \Upsilon_{\min}) \le K_{suf}\left(\deg_i + \deg_j\right)$, frequency synchronization will be achieved in the network. End of proof of Statement 1.

Proof of Statement 2: In a synchronized network, Eq. (20) holds for all N_G generators. Applying trigonometric transformations and summing both sides of the equation for N_G, we will have $\omega_{syn}\sum_{i=1}^{N_G} D_i$ on the left side. On the right side, due to the oddness of the sine function and the symmetry of phase shifts between agents, the expression $\sum_{i=1}^{N_G}\omega_i - \sum_{i=1}^{N_G} X_i$ is obtained. So, the actual frequency of the network will be given by Eq. (22). End of proof of Statement 2.

Result 1: In a network with losses or non-zero phase shift in the dynamic model, the angle ϕ_{ij} leads to network oscillation at a frequency other than the nominal frequency ω. Through constant input power to the oscillatory agents while changing the phase shift ϕ_{ij}, the value of the network oscillation frequency will change as shown in **Figure 1**.

Result 2: For a network with homogeneous agents $D_i = D_j = D$ and uniform connectivity strength given by $P_{ij} = P/N_G$, the necessary condition for synchronization is derived as $P \ge \frac{N_G(\omega_{\max} - \omega_{\min})}{2(N_G - 1)\sin(\rho + \phi_{\max})}$. For a lossless network $\phi_{max} = 0$ and $\rho = \pi/2$, this condition simplifies to $P \ge \frac{N_G(\omega_{\max} - \omega_{\min})}{2(N_G - 1)}$, which is equivalent to the condition presented in [33] and Theorem 2.1 in Ref. [12]. This condition further reduces to $P \ge (\omega_{\max} - \omega_{\min})$ for $N_G = 2$ and $P \ge \frac{(\omega_{\max} - \omega_{\min})}{2}$ as $N_G \to \infty$ approaches infinity.

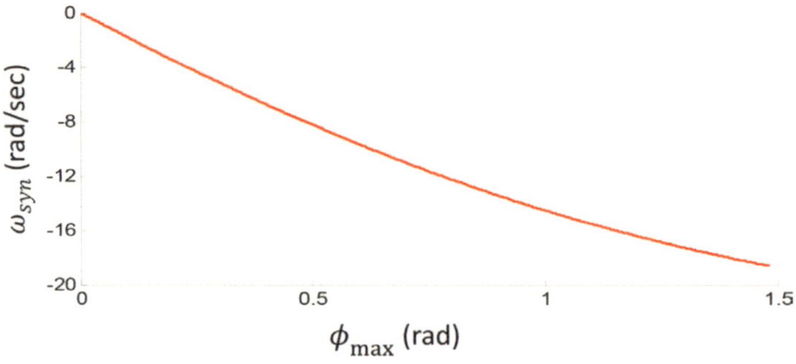

Figure 1.
Effect of phase shift on the rotational frequency of agents in the second-order non-uniform Kuramoto model.

Additionally, assuming zero phase shift, the condition becomes $P \geq \frac{N_G(\omega_{\max}-\omega_{\min})}{2(N_G-1)\sin(\rho)}$, which is more conservative compared to the result presented in Theorem 3.1 in Ref. [13].

3.2 Sufficient condition for synchronization - LaSalle invariance principle

Theorem 2: Consider the power network oscillators' dynamic model (11), where the coupling strength $P = P^T$ *and* $B = H^T \in \mathbb{R}^{N_G \times |\varepsilon|}$ *is the incidence matrix of the complete graph. For* $\rho \in \left[0, \pi - \left(H\overline{\theta}_e + 2H\overline{\vartheta}\right)\right]$*, let the coupling topology between oscillators be such that the algebraic connectivity of the network graph* $\lambda_2\left(L\left(P_{ij}\cos\left(\vartheta_{ij}\right)\right)\right)$ *exceeds the* $\lambda_2^{Critical1}$ *or* $\lambda_2^{Critical2}$*, which are defined as:*

$$\lambda_2\left(L\left(P_{ij}\cos\left(\vartheta_{ij}\right)\right)\right) \geq \lambda_2^{Critical1} = \frac{\max_{i\neq j}\{D_iD_j\}\left\|HD^{-1}\omega\right\|_2 + d\left\|\mathrm{diag}\left(P_{ij}\sin\left(|\vartheta_{ij}|\right)\right)\mathbf{1}_{|\varepsilon|}\right\|_2}{\kappa\rho(d/N_G)\mathrm{sinc}(\rho)} \tag{23}$$

$$\lambda_2\left(L\left(P_{ij}\cos\left(\vartheta_{ij}\right)\right)\right) \geq \lambda_2^{Critical2} = \frac{\max_{i\neq j}\{D_iD_j\}\left\|HD^{-1}\omega\right\|_2 K_{\max}}{\kappa\rho(d/N_G)\mathrm{sinc}(\rho)} \tag{24}$$

Here, $K_{\max} = \max_{i\neq j}\left\{\mathbf{1}_{|\varepsilon|} - \mathrm{diag}\left(P_{ij}/P_{ij}^0\right)\cos(\rho)\right\}$, θ_e is equal with network equilibrium point, $H\overline{\theta}_e = \max_{i\neq j}\{|\theta_{e,i} - \theta_{e,j}|\}$, $H\overline{\Theta}^0 = \max_{i\neq j}\left\{|\Theta_i^0 - \Theta_j^0|\right\}$,
$H\overline{\vartheta} = H\overline{\Theta}^0 + \phi_{\max}$, P_{ij}^0 is equal with network coupling strength before the transient, $0 \leq \kappa \leq Q$, $d: = \sum_{i=1}^{N_G} D_i$, $m: = \sum_{i=1}^{N_G} M_i$, $\vartheta_{ij} = \Theta_{ij}^0 + \phi_{ij}$, Θ_{ij}^0 represents the initial phase difference of the generators, $Q = \sqrt{\lambda_U^{\min}/\left(\lambda_U^{\max} + \mathcal{S}_{\max}\right)}$,
$\mathcal{S}_{\max} = (\alpha + \beta)(m/2)\max_{i\neq j}\{P_{ij}\}$, $\phi_{\max} = \max_{i\neq j}\left\{\phi_{ij}\right\}$ *and* $\lambda_U^{\max}$ *and* $\lambda_U^{\min}$ are respectively the largest and smallest eigenvalues of the matrix U with the following relationship:

$$U = \begin{bmatrix} \mathrm{diag}(D_iD_j) & \mathrm{diag}(D_iD_j)HD^{-1}M \\ MD^{-1}H^T\mathrm{diag}(D_iD_j) & \alpha H^T\mathrm{diag}(M_iM_j)H + \beta m M \end{bmatrix} \tag{25}$$

And α and β are also satisfied in the $\alpha > \max_{i \neq j}\{D_j M_i / D_i M_j\}$ *and* $\beta \geq (d/m)\max\{M_i/D_i\} + (\alpha - 1)(N_G - 2)\max\{M_i\}/m - \alpha$. *(Then)*

Statement 1: Phase Cohesiveness. The set $\Omega_\varsigma = \{(H\theta, \dot{\theta}) \in \mathbb{R}^{|\varepsilon| \times N_G} \ : \ \|H\theta(t)\|_2 \leq (\kappa/Q)\rho\}$ is a positively invariant set for the electrical network, and under the given conditions, if the initial phase difference of the network generators satisfies $\|H\theta(t_0)\| = \kappa\rho \leq Q\rho$, the network states will be uniformly bounded and the phase difference of the generators will remain within the region $\|H\theta(t)\| \leq (\kappa/Q)\rho$.

Statement 2: Frequency Synchronization. Under the initial conditions mentioned above, the states of the oscillatory components in the dynamic model of the electrical network converge towards the largest positively invariant set given by $S_{inv} = (H\theta(t), \dot{\theta}) = (H\theta_e(t), \boldsymbol{\omega}_{syn})$, which is equivalent to achieving frequency synchronization in the electrical network in the form $\lim_{t \to \infty} |\dot{\theta}_i(t) - \dot{\theta}_j(t)| = 0, \ \forall \ i \neq j$, *or* $\|H\dot{\theta}(t)\|_2 = 0$. *In this set,* $\boldsymbol{\omega}_{syn} = \mathbf{1}_{N_G}\omega_{syn}n$, *and* ω_{syn} *is obtained from Eq. (22). The largest positively invariant set with the equilibrium point of the system* $(H\theta(t), \dot{\theta}) = (H\theta_e(t), \boldsymbol{\omega}_{syn})$ *is unique.*

Proof: To prove this theorem, the Lyapunov function is defined as follows. The details of its proof are provided in [82, 83].

$$E_1(H\theta, H\dot{\theta}) = \frac{1}{2}\begin{bmatrix}(H\theta)^T & \dot{\theta}^T\end{bmatrix}\begin{bmatrix}\text{diag}(D_iD_j) & \text{diag}(D_iD_j)HD^{-1}M \\ MD^{-1}H^T\text{diag}(D_iD_j) & \alpha H^T\text{diag}(M_iM_j)H + \beta mM\end{bmatrix}\begin{bmatrix}H\theta \\ \dot{\theta}\end{bmatrix}$$
$$E_2(H\theta) = (\alpha + \beta)\int_0^{H\theta}\text{diag}(M_iM_j)\left[-HM^{-1}\omega + HM^{-1}H^T\text{diag}(P_{ij})\,\mathbf{sin}\,(H\theta + H\vartheta))\right]d(H\theta) \tag{26}$$

Result 3 (The region of Attraction Development with Input Power Control): According to (23) and (24), it is evident that reducing the term $D_iD_j|\omega_i/D_i - \omega_j/D_j|$ *in the edges of the electrical network graph leads to a decrease in the term* $\max_{i \neq j}\{D_iD_j\}\|HD^{-1}\omega\|_2$ *and ultimately a decrease in* $\lambda_2^{Critical1}$ *and* $\lambda_2^{Critical2}$. *The term* ω_i *in the electrical network is defined as* $\omega_i = P_{m,i} - P_{L2G,i}$, *where* $P_{m,i}$ *represents the mechanical power input to generator i and* $P_{L2G,i}$ *represents the influence of network loads on generator i. Therefore, applying an appropriate control strategy to regulate parameters* $P_{m,i}$ *and* $P_{L2G,i}$ *in the network can provide a sufficient condition for synchronization.* **Figure 2** illustrates the

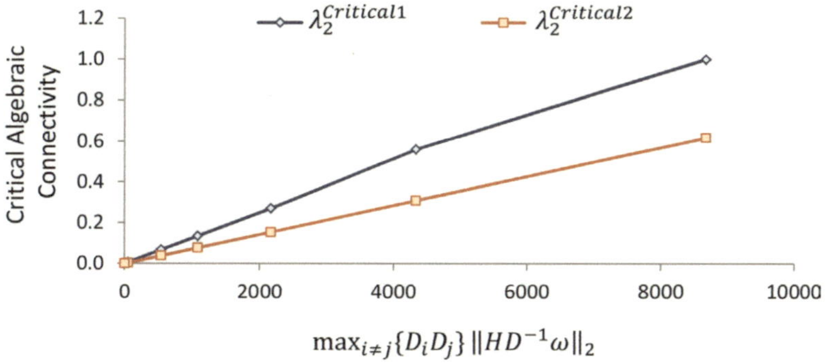

Figure 2.
Algebraic connectivity as a function of $\max_{i \neq j}\{D_iD_j\}\|HD^{-1}\omega\|_2$.

variations of $\lambda_2^{Critical1}$ and $\lambda_2^{Critical2}$ with respect to changes in the term $max_{i \neq j}\{D_i D_j\}\|HD^{-1}\omega\|_2$.

Result 4 (Impact of Oscillators' Initial Phase on the Critical Coupling): According to the results, as the initial phase difference Θ_{ij}^0 increases towards $\pi/2$, the value of the term $\cos(\vartheta_{ij})$ decreases and the value of the term $\sin(|\vartheta_{ij}|)$ increases. Consequently, the values of $\lambda_2^{Critical1}$ and $\lambda_2^{Critical2}\lambda$ decrease. Considering the above, in case of reducing the fault clearance time in the network, the stability conditions and the region of the attraction of the network will improve.

Result 5 (The region of attraction Development with Control of Agent Parameters): As the inertia and damping coefficient of generators in the electrical network become closer to a homogeneous state, the value of critical coupling decreases. In the extreme case where the inertia and damping coefficient are the same for all oscillatory agents, we have:

$$\lambda_2^{Critical1} = \frac{\|H\omega\|_2 + N_G\|\mathrm{diag}(P_{ij}\sin(|\vartheta_{ij}|))\mathbf{1}_{|\varepsilon|}\|_2}{\kappa\rho\mathrm{sinc}(\rho)} \quad (27)$$

$$\lambda_2^{Critical2} = \frac{\|H\omega\|_2 K_{\max}}{\kappa\rho\mathrm{sinc}(\rho)} \quad (28)$$

In this case, the value of critical coupling is not dependent on the parameters of the generators in the network. **Figure 3** illustrates the variations of $\lambda_2^{Critical1}$ and $\lambda_2^{Critical2}$ for two homogeneous and heterogeneous networks with respect to changes in ρ. It is evident that the value of critical coupling in the homogeneous network is smaller than that in the heterogeneous network.

Result 6 (The effect of transmission line resistance in the dynamic of an electrical network): Employing trigonometric transformation, it is evident that the phase shift ϕ_{ij} in the electrical network model not only affects the magnitude of the connections between network generators from P_{ij} to $P_{ij}\cos(\phi_{ij})$, but also adds the term $-\sum_{j=1}^{N_G} P_{ij}\sin(\phi_{ij})\cos(\Theta_i - \Theta_j) < 0$, to the ith generator node. Both terms have a negative impact on the stability conditions of the network. Therefore, it can be concluded that

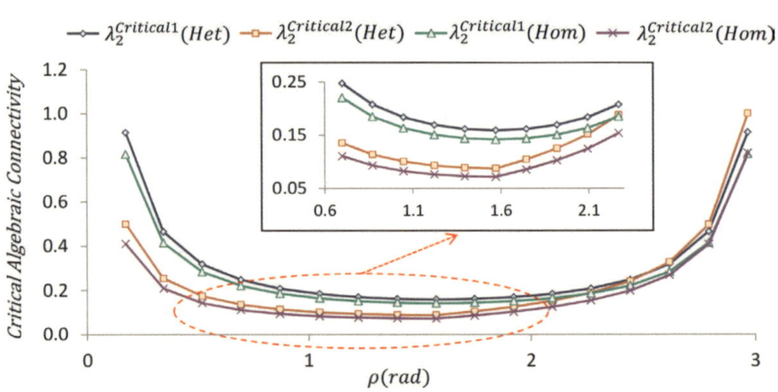

Figure 3.
Algebraic connectivity in two "heterogeneous" and "homogenous" networks (abbreviated to "Het" and "Hom", respectively).

Stage	Function
1	Obtain input information for each bus in the network, including specifications of network generators, transmission line impedances, load specifications, transformer specifications, and convert them into an appropriate format for AC load flow analysis.
2	Perform AC load flow analysis and determine the values of network's state
3	Apply the Kron reduction technique and develop the dynamic model of the network with a complete graph.
4	Develop network topology in pre-fault, during fault, and post-fault conditions based on the postulated scenario.
5	Determine the time step size $t_s = 0.001^*$ *sec* for numerical simulations and set the fault removal time as $t_c = t_s$.
6	Perform simulations for the state equations from the time of fault occurrence until fault removal time t_c to calculate initial values of state variables.
7	Calculate initial conditions for network state variables and calculate parameters Q, $\mu_{c1} = \lambda_2^{Critical1}$, $\mu_{c2} = \lambda_2^{Critical2}$, $\lambda_2(L(P_{ij}\cos(\vartheta_{ij})))$, and determine ρ at the fault clearing time.
8	If $\mu_{c1} \leq \|H\theta(t_0)\|_2 \leq Q\rho$ or $\mu_{c2} \leq \|H\theta(t_0)\|_2 \leq Q\rho$, then the network state variables are in the region of attraction at the fault clearing time. The network maintains its transient stability and proceed to step 9. If $\|H\theta(t_0)\|_2 \geq Q\rho$, the network cannot maintain stability condition at the fault removal time t_c. However, the network has maintained transient stability condition in the previous step, and the critical clearing time or CCT is equal to $t_{cr} = t_c - t_s$. Proceed to step 10.
9	Update the fault removal time as $t_c = t_c + t_s$ *sec* sec and go back to step 6.
10	Determine the value of CCT as t_{cr} in the output.

* Since the period of the first swing angle of the rotor generator is approximately 2–4 seconds, a time step of $t_s = 0.001$ has been chosen.

Table 1.
Estimating the region of attraction and critical clearing time t_c based on the initial condition.

increasing the resistance of transmission lines makes it more difficult to establish the sufficient condition for phase coherence among network generators.

Table 1 presents a recursive algorithm for estimating the critical clearing time t_c under conditions where an error occurs in the electrical network topology.

3.3 Sufficient condition for synchronization –contraction property

The challenges in finding a Lyapunov function and applying the Lyapunov method to analyze the stability of certain systems, especially nonlinear multi-agent systems with time-varying communications, demonstrate the need for more advanced tools and methods in studying these systems. In [84–92], several examples of these tools have been developed in the form of practical theories for analyzing the stability properties of nonlinear monotone systems using max-separable Lyapunov functions. Conditions for convergence of states in multi-agent systems with nonlinear dynamics have been developed using analytical theories in the field of non-smooth analysis. In this section, we invoke the contraction property introduced in [93] for consensus protocols in self-organizing systems. According to the mentioned theory, in order to achieve convergence of agent states to a common value, it is necessary for $\max\{x_1, \dots, x_n\}$ to be a non-increasing function of time and at the same time, $\min\{x_1, \dots, x_n\}$ should not decrease. Therefore, by combining these two characteristics in [93], the Lyapunov function $V(x) = \max\{x_1, \ldots, x_n\} - \min\{x_1, \ldots, x_n\}$ or in another form, $V(x) = \max\{|x_i - x_j| \quad i|j \in \{1, \ldots, n\},\}$ has been introduced to analyze the stability of the consensus protocol. Since the max-separable Lyapunov function is not necessarily continuously differentiable, the direct Lyapunov method cannot be applied to this class of Lyapunov functions.

In [84], with reference to [94, 95], it has been shown that for a system $\dot{x} = f(t,x),\ f : \mathbb{R} \times \mathcal{D} \to \mathbb{R}^n$ where $\mathcal{D} \subset \mathbb{R}^n$ is a set and $V(t,x) : \mathbb{R} \times \mathcal{D} \to \mathbb{R}$ is a continuous function that satisfies the Lipschitz condition, the *Upper Right Dini Derivative* bound of the function $V(t,x)$ along the state trajectories $f(t,x)$ can be presented as the $D_f^+ V$ function, as follows:

$$D_f^+ V(t,x) = \limsup_{\tau \to 0^+} \frac{V(t+\tau, x(t,\tau)) - V(t,x)}{\tau} = \limsup_{\tau \to 0^+} \frac{V(t+\tau, x+\tau f(t,x)) - V(t,x)}{\tau} \tag{29}$$

In this case, considering $V_i(t,x(t)) : \mathbb{R} \times \mathcal{D} \to \mathbb{R}$ as a class C^1 function and defining $V(t,x(t)) = \max_{i \in \{1, \ldots, n\}} V_i(t,x(t))$, the equality $D^+ V(t,x(t)) = \max_{i \in \{1, \ldots, n\}} \dot{V}_i(t,x(t))$ holds. The stability and control analysis theories derived from this approach can be applied in areas related to distributed decision-making, coordination, synchronization, and coherence of states in multi-agent systems and high-dimensional systems. The development of Theorem 3 is based on this approach.

Lemma 2: We recall the dynamic model of a lossy power network (19), where the network connectivity matrix $P = P^T$ is a complete graph, the symmetric phase shift ϕ_{ij} and phase Θ_i are restricted to the regions $\phi_{ij} \in (0, \pi/2)$ and $\{\Theta(t) \in \Delta(\rho) : \rho \in (0, \pi/2)\}$, and the network coupling matrix is denoted as $B = H^T \in \mathbb{R}^{N_G \times |\varepsilon|}$. The upper bound values for the oscillators' frequency and its rate of change are given by $\dot{\theta}_{\max} = \omega_{\square \times Dif} + c$ *and* $\ddot{\theta}_{\max} = \left(\frac{D_{\min} + D_{\max}}{M_{\min}}\right) \omega_{\square \times Dif} + \frac{D_{\max}}{M_{\min}} c$, *where* $\omega_{syn}^0 = \left(\sum_{i=1}^{N_G} \omega_i - S_\omega^0\right) / \sum_{i=1}^{N_G} D_i$, $S_\omega^0 = \sum_{i=1}^{N_G} X_i^0$, $X_i^0 = \sum_{j=1, j \neq i}^{N_G} P_{ij}^0 \sin\left(\phi_{ij}\right) \cos\left(\Theta_i^0 - \Theta_j^0\right)$, $\omega_{Dif} = \left(\omega_{\max} + \sin(H\overline{\vartheta}) \deg_{\max}\right) / D_{\min}$, $c \geq \|\omega_{Dif}| - |\omega_{syn}^0\|$, $H\overline{\vartheta} = H\overline{\Theta}^0 + \phi_{\max}$, *and* $H\overline{\Theta}^0 = \max_{i \neq j}\left\{|\Theta_i^0 - \Theta_j^0|\right\}$. Additionally, Θ_i^0 and P_{ij}^0 represent the initial phase of agent i and the initial strength of communication between agents i and j at time $t = 0$, (respectively)

Proof: The proof of this lemma is provided in [96].

In continuation of the studies [12] (Theorem 4.1), [13] (Theorem 3.1), [27], [33] (Theorem 2), [34, 68–71], this section provides a sufficient condition for achieving synchrony in a lossy electrical network with heterogeneous oscillators, non-uniform inter-agent connectivity matrix. The results of this theorem also hold for networks with uncertainty in the agents' input power in the range $\omega_i \in \{\omega_{\min} - \omega_{\max}\}$.

Theorem 3: *We consider the dynamic model of an electrical network with non-zero phase shifts Eq. (17). If the following condition is satisfied in the network graph:*

$$N_G \min_{i \neq j} \left\{\frac{P_{ij}}{D_i} \cos\left(\phi_{ij}\right)\right\} \sin(\rho) \geq \max_{i \neq j} \left|\frac{\omega_i}{D_i} - \frac{\omega_j}{D_j}\right| + 2\ddot{\Theta}_{\max} \max_{i \in \{1, \ldots N_G\}} \left\{\frac{M_i}{D_i}\right\} + N_G \max_{i \neq j} \left\{\frac{P_{ij}}{D_i} \sin\left(\phi_{ij}\right)\right\} - N_G \min_{i \neq j} \left\{\frac{P_{ij}}{D_i} \sin\left(\phi_{ij}\right)\right\} \cos(\rho) \tag{30}$$

Then, the set $\Theta \in \Delta(\rho)$ is an attractive region for the electrical network with respect to $\rho \in [0, \pi/2 - \phi_{\max}]$, and under the condition that the initial phases of the agents fall within this non-empty set, phase coherence will be established in the network.

Proof: To prove this Theorem, the max-separable Lyapunov function $V(\Theta(t)) = \max_{i,j \in \{1, \dots, N_G\}} V_{ij}(\Theta(t)) = \Theta_p(t) - \Theta_q(t)$ is used. The phases Θ_p and Θ_q correspond to the oscillators Osc_p and Osc_q at the boundary positions of the arc ρ, as shown in **Figure 4**. In order for $\Delta(\rho)$ to be invariant, it is necessary that the geodesic distance between the two boundary agents decreases. This distance is equal to the maximum $V_{ij}(\Theta(t))$, which is why we define the candidate Lyapunov function $V(\Theta(t)) = \max_{i,j \in \{1, \dots, N_G\}} V_{ij}(\Theta(t)) = \Theta_p(t) - \Theta_q(t)$ on $\mathbb{T}^1$. The function $V(\Theta(t))$ is not continuously differentiable, but it satisfies the Lipschitz continuity conditions. Therefore, according to the Lemma 2.2 in [84], to ensure the decremental nature of $V(\Theta(t))$, it is necessary that the upper bound of the Dini Derivative of $V(\Theta(t))$ along the trajectory of the system Eq. (17) is negative.

$$D_f^+ V(\Theta(t)) = \limsup_{\tau \to 0^+} \frac{V(\Theta(t+\tau)) - V(\Theta(t))}{\tau} = \dot{\Theta}_p(t) - \dot{\Theta}_q(t) < 0 \tag{31}$$

The details of the proof of this Theorem are presented in Ref. [96].

Result 7 (The region of attractive expansion through power control): Considering the sufficient condition for phase coherence of the agents given in Eq. (30), it is clear that reducing the differential term $\max_{i \neq j} |\omega_i/D_i - \omega_j/D_j|$ in the edges of the graph associated with the dynamic model of the electrical network leads to a decrease in the right-hand side of the Eq. (30). In other words, the heterogeneity of ω_i/D_i in each node of the network will have a negative impact on the stability condition. **Figure 5** illustrates the effect of ω_i/D_i heterogeneity on the critical angle ϕ_{cri} (for a network with a fixed topology, the values of the left-hand side and the right-hand side of Eq. (30) will be

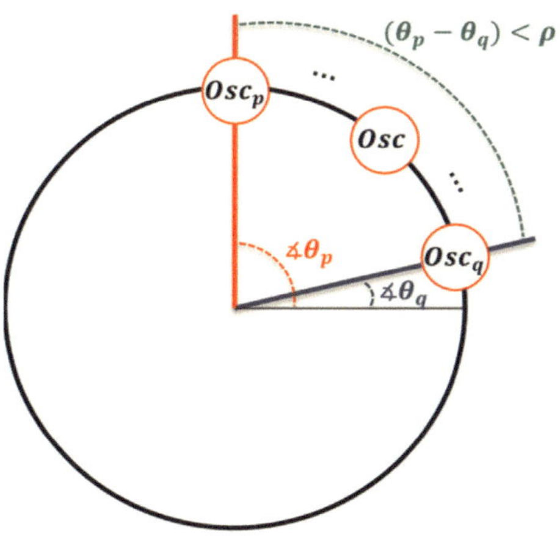

Figure 4.
Network agents' positions in an extreme state.

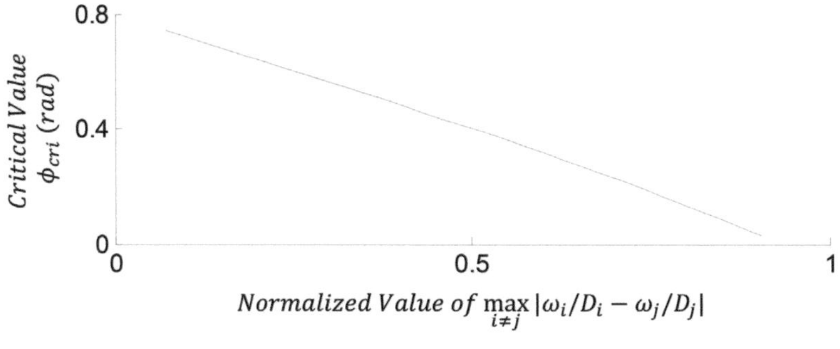

Figure 5.
The value of ϕ_{cri} as a function of $\max_{i\neq j} |\omega_i/D_i - \omega_j/D_j|$.

equal if the phase shift in the coupling function is equal to ϕ_{cri}) in a sample network. According to Eq. (17), the value of ω_i in the dynamic model of the electrical network is given by $\omega_i = P_{m,i} - P_{L2G,i}$, where $P_{m,i}$ represents the mechanical power input to the i-th generator and $P_{L2G,i}$ represents the impact of network loads on the i-th generator after applying the Kron reduction method. Thanks to the control systems in the electrical network, the value of the mechanical input power and partially the power consumption at the network loads are adjustable. By implementing appropriate control strategies to regulate these parameters, it is possible to make the synchronization condition stated in Eq. (30) more achievable.

Result 8: The term $2\ddot{\Theta}_{\max} \max_{i\in\{1,\dots,N_G\}} \{M_i/D_i\}$ on the right-hand side of (30) indicates that for two networks with the same network graph and generators' parameters M_i and D_i being pairwise equal, it will be easier to satisfy the phase coherence condition of the agents in the network with a smaller $\ddot{\Theta}_{max}$.

Result 9: As the inertia and damping coefficients of the generators approach a homogeneous state, the magnitudes of the terms $\min_{i\neq j} \left\{P_{ij}/D_i \sin\left(\phi_{ij}\right)\right\} \cos(\rho)$ and $\max_{i\neq j} \left\{P_{ij}/D_i \sin\left(\phi_{ij}\right)\right\}$ become closer to each other, and the value of $\max_{i\neq j} \left|\omega_i/D_i - \omega_j/D_j\right|$ also decreases. Considering this aspects, having a large standard deviation of the generator parameters reduces the right-hand side of the inequality Eq. (30). In the extreme case where the inertia and damping coefficients of all network generators are the same and $\rho = \pi/2 - \phi_{\max}$, the condition will simplifies to:

$$N_G \min_{i\neq j} \left\{P_{ij} \cos\left(\phi_{ij}\right)\right\} \cos(\phi_{\max}) \geq \max_{i\neq j} \left|\omega_i - \omega_j\right| + 2M\ddot{\Theta}_{\max} + N_G \sin(\phi_{\max}) \left(\max_{i\neq j} \{P_{ij}\} - \min_{i\neq j} \left\{P_{ij} \sin\left(\phi_{ij}\right)\right\}\right) \quad (32)$$

Result 10: The closer the coupling strength between network generators becomes homogeneous, it will be easier to achieve the sufficient condition for synchronization. In the extreme case, for a weakened network with homogeneous connections and parameters and $\rho = \pi/2 - \phi$, *the sufficient condition will be as follows:*

$$P \geq \frac{\max_{i\neq j} \left|\frac{\omega_i}{D_i} - \frac{\omega_j}{D_j}\right| + 2\ddot{\Theta}_{\max} \max_{i\in\{1,\dots,N_G\}} \left\{\frac{M_i}{D_i}\right\}}{\min_{i\neq j} \{1/D_i\} - \max_{i\neq j} \{\sin(\phi)/D_i\}} \quad (33)$$

Furthermore, if the desired network also consists of homogeneous oscillators, the condition simplifies to:

$$P \geq \frac{\max_{i \neq j} |\omega_i - \omega_j| + 2\ddot{\Theta}_{\max} M}{1 - \sin(\phi)} \tag{34}$$

Result 11: The arc consisting the phase of agents, $\rho \in [0, \pi/2 - \phi_{\max}]$, is introduced in the form of two terms $\sin(\rho)$ and $\cos(\rho)$ in the inequality Eq. (30). It is evident that as the value of ρ increases towards $\pi/2 - \phi_{\max}$, the right-hand side of the inequality decreases, while the left-hand side increases. In the extreme case of $\rho = \pi/2 - \phi_{\max}$, it becomes easier to satisfy the sufficient condition, and it transforms into the following equation:

$$\begin{aligned} N_G \min_{i \neq j} \left\{ \frac{P_{ij}}{D_i} \cos\left(\phi_{ij}\right) \right\} \cos(\phi_{\max}) \geq & \max_{i \neq j} \left| \frac{\omega_i}{D_i} - \frac{\omega_j}{D_j} \right| + 2\ddot{\Theta}_{\max} \max_{i \in \{1, \dots N_G\}} \left\{ \frac{M_i}{D_i} \right\} \\ & + N_G \max_{i \neq j} \left\{ \frac{P_{ij}}{D_i} \sin\left(\phi_{ij}\right) \right\} \\ & - N_G \min_{i \neq j} \left\{ \frac{P_{ij}}{D_i} \sin\left(\phi_{ij}\right) \right\} \sin(\phi_{\max}) \end{aligned} \tag{35}$$

Result 12: According to Eq. (30), which shows the impact of transmission line resistance in the form of phase shift ϕ_{ij}, since the term $N_G \max_{i \neq j} \left\{ P_{ij}/D_i \sin\left(\phi_{ij}\right) \right\} - N_G \min_{i \neq j} \left\{ P_{ij}/D_i \sin\left(\phi_{ij}\right) \right\} \cos(\rho)$ is always greater than zero, increasing the resistance of transmission lines leads to an increase in the value of phase shift ϕ_{ij}, making it more difficult to achieve the sufficient condition for the phase coherence of network generators. In the case where the resistance value is zero, condition in Eq. (30) simplifies to the following equation:

$$N_G \min_{i \neq j} \left\{ \frac{P_{ij}}{D_i} \right\} \sin(\rho) \geq \max_{i \neq j} \left| \frac{\omega_i}{D_i} - \frac{\omega_j}{D_j} \right| + 2\ddot{\Theta}_{\max} \max_{i \in \{1, \dots, N_G\}} \left\{ \frac{M_i}{D_i} \right\} \tag{36}$$

Result 13: In the networks of oscillators with the first order model $M = 0$, homogenous agents $D_i = D_j$, and uniform connections $P_{ij} = P/N_G$, when the phase shift ϕ_{ij} is equal to zero, the condition Eq. (30) is equivalent to the condition of Theorem 3.1 in [13] and is expressed as $P \geq (\omega_{\max} - \omega_{\min}) / \sin(\rho)$.

4. Conclusion

In this chapter, we present the estimation of the region of attraction in an electric network by applying the LaSalle invariance principle and the contraction property for consensus protocols in systems with autonomous agents. We detail this estimation in Theorem 2 and 3. Furthermore, Theorem 1 introduces the necessary condition for achieving synchronization based on the network graph and its parameters.

In Theorem 2, we explicitly express the estimation of the region of attraction of the electric network as a function of the underlying network parameters, the input power intensity of the network agents, the intensity of the coupling topology, and the initial conditions of the model. This estimation is presented in two distinct forms:

- The region of attraction based on the topology of network graph $\lambda_2(L(P_{ij}\cos(\vartheta_{ij}))) \geq \lambda_2^{Critical}$.
- The region of attraction based on the network's initial conditions $\mu_c \leq \|H\theta(t_0)\|_2 \leq Q\rho$.

Furthermore, based on the results of Theorem 2, an algorithm for estimating the critical clearing time in the network is presented.

In Theorem 3, the obtained estimation for the region of attraction of the electric network is presented as an invariant set of agent phases and also an explicit function of the underlying network parameters, input power intensity of the network agents, the intensity of their interactions, and the initial conditions of the model.

Based on the results obtained from theorems 2 and 3, it is necessary to increase the distance between the value of $\lambda_2(L(P_{ij}\cos(\vartheta_{ij})))$ and the critical value in order to develop the estimated regions of attractions. This can be achieved by keeping $\lambda_2(L(P_{ij}\cos(\vartheta_{ij})))$ constant and reducing $\lambda_2^{Critical}$, or by keeping $\lambda_2^{Critical}$ constant and increasing $\lambda_2(L(P_{ij}\cos(\vartheta_{ij})))$. The development of the region of attraction is introduced in the following methods:

1. Increasing the connectivity intensity between the generators of the electric network, which consequently increases the value of the connectivity distance of the network, $\lambda_2\left(L\left(P_{ij}\cos\left(\phi_{ij}\right)\right)\right)$.
2. Increasing the density of connections in the electric network (pre-reduction Kron topology). As a result of this process, the value of the algebraic connectivity of the network, $\lambda_2\left(L\left(P_{ij}\cos\left(\phi_{ij}\right)\right)\right)$ (increases)
3. Having the term $D_iD_j|\omega_i/D_i - \omega_j/D_j|$ in the network generators $i \in \{1, \dots, N_G\}$ close to each other (details in the result of 6–2).
4. The inertia and damping coefficients of the generators should be closer to a state of homogeneity.

Based on the details discussed in this chapter, and the analytical presented theories, the following topics can be suggested for further exploration within the scope of this chapter. These topics are based on providing analytical results for second-order Kuramoto oscillator networks under different operating conditions and improving the capabilities of electric networks to maintain stability.

4.1 Further network analysis with more details

- Investigation and analysis of the impact of uncertainty in the parameters of oscillatory agents (damping, inertia, input power or natural frequency, and the intensity of inter-agent connection) on the stability conditions addressed in this chapter for second-order Kuramoto oscillator networks.

- Examination and analysis of the effect of the delay in the second-order Kuramoto oscillator model on the stability condition and identification of the permissible delay before the model reaches an unstable state (necessary and sufficient conditions for stability in networks with phase delay should be provided).

4.2 Moving from network analysis towards control theory formulation

- Enhancement of the region of attraction and transient response in second-order Kuramoto oscillator networks through the design of optimized controllers.

- Improvement of the estimated the region of attraction and transient response in second-order Kuramoto oscillator networks with time-varying connecting topology through the design of adaptive controllers to control the input power to the oscillators.

- Controller design for electric networks to improve the region of attraction and transient response of the network by mapping the network model to a second-order Kuramoto oscillator model and identifying the critical clearing time based on the derived stability condition.

- Analysis of the impact of delay or pocket loss in the communication feedback on the stability conditions of electric networks.

4.3 Numerical analysis

- Performing numerical analyses and investigating the impact of oscillatory agent parameters and inter-agent connections on the stability conditions of oscillatory networks.

Author details

Javad Soleimani[1], Reza Farhangi[1*], Gunes Karabulut Kurt[1] and Fatemeh Mechershavi[2]

1 Department of Electrical Engineering , Poly-Grames Research Center, Polytechnique Montreal, Montreal, Quebec, Canada

2 Energy Engineering Department, Sharif University of Technology, Tehran, Iran

*Address all correspondence to: farhangi.reza@gmail.com

References

[1] Kuramoto Y. Self-entrainment of a population of coupled non-linear oscillators. International Symposium on Mathematical Problems Theoretical Physics. 1975:420-422

[2] Michel AN, Fouad A, Vittal V. Power system transient stability using individual machine energy functions. Circuits and Systems, IEEE Transactions on. 1983;**30**(5):266-276

[3] Chiang H-D, Wu FF, Varaiya PP. Foundations of the potential energy boundary surface method for power system transient stability analysis. IEEE Transactions on Circuits and Systems. 1993;**35**:712-728

[4] Chiang H-D, Chu CC. Theoretical foundation of the BCU method for direct stability analysis of network-reduction power system models with small transfer conductance. IEEE Transactions on Circuits and Systems. 1995;**42**:252-265

[5] Padiyar KR, Ghosh KK. Dynamic security assessment of power systems using structure-preserving energy functions. International Journal of Electrical Power & Energy Systems. 1989;**11**:39-46

[6] Alberto LFC, Silva FHJR, Bretas NG. Direct methods for transient stability analysis in power systems: state of art and future perspectives. In: IEEE on Power Tech Proceedings. 2001

[7] Chiang H-D. Direct Methods for Stability Analysis of Electric Power Systems: Theoretical Foundation, BCU Methodologies, and Applications. Hoboken, NJ, USA: John Wiley & Sons; 2011

[8] Long T, Turitsyn K. Synchronization Stability of Lossy and Uncertain Power Grids. American Control Conference (ACC). 2015. pp. 5056-5061

[9] Chiang H-D. Study of the existence of energy functions for power systems with losses. Circuits System, IEEE Transactions on. 1989;**39**(11): 1423-1429

[10] Hill DJ, Chen G. Power systems as dynamic networks. In: IEEE International Symposium on Circuits and Systems. Kos, Greece; 2006. pp. 725-722

[11] Jadbabaie A, Motee N, Barahona M. On the stability of the Kuramoto model of coupled nonlinear oscillators. In: Proceedings on American Control and Conference. 2005. pp. 4296-4301

[12] Chopra N, Spong MW. On exponential synchronization of Kuramoto oscillators. IEEE Transactions on Automatic Control. 2009;**54**:353-357

[13] Choi YP, Ha SY, Jung S, Kim Y. Asymptotic formation and orbital stability of phase-locked states for the Kuramoto model. Physica D. 2012; **241**(7):735-754

[14] Ha SY, Kim Y, Li Z. Asymptotic Synchronous Behavior of Kuramoto Type Models With Frustrations. Networks and Heterogenious Media. 2014;**9**(1):33-64

[15] Levnajic Z. introduced the notion of link frustration to characterize and quantify the dynamical states of networks. Physical Review E. 2011;**84**(1)

[16] Zheng ZG. Frustration effect on synchronization and chaos in coupled oscillators. Chinese Physical Society. 2001;**10**(8)

[17] Choi C, Oh E, Kahng B, Kim D. Modular synchronization in complex networks with a gauge Kuramoto model. In: International Conference on Complex Sciences. Vol. 4. 2009

[18] Park K, Rhee SW, Choi MY. Glass synchronization in the network of oscillators with random phase shift. Physical Review E. 1998;**57**:5030-5035

[19] Tanaka T, Aoki T, Aoyagi T. Dynamics in co-evolving networks of active elements. Forma. 2009;**24**:17-22

[20] Levnajic Z. Emergent multistability and frustration in phase-repulsive networks of oscillators. Physical Review E. 2011;**84**

[21] Filatrella G, Nielsen AH, Pedersen NF. Analysis of a power grid using a Kuramoto-like model. European Physical Journal B. 2008;**61**(4):485-491

[22] Fortuna L, Frasca M, Fiore AS. Analysis of the Italian power grid based on Kuramoto-like model. In: 5th International Conference on Physics and Control, Leon, Spain. Sep 2011

[23] Dörfler F, Bullo F. Synchronization of power networks: Network reduction and effective resistance. In: IFAC Workshop on Distributed Estimation and Control in Networked Systems. Annecy, France; 2010. pp. 197-202

[24] Dörfler F, Bullo F. Topological equivalence of a structure-preserving power network model and a non-uniform kuramoto model of coupled oscillators. In: 50th IEEE conference on decision and control and European control conference, Orlando, FL, USA. 2011

[25] Colombino M, Groß D, Brouillon J-S, Dörfler F. Global phase and magnitude synchronization of coupled oscillators with application to the control of grid-forming power inverters. IEEE Transactions on Automatic Control. 2019;**64**(11)

[26] Dörfler F, Bullo F. Exploring synchronization in complex oscillator networks. In: Proceedings of 51 th IEEE Conference on Decision and Control (CDC). Maui; 2012. pp. 7157-7170

[27] Dörfler F, Bullo F. Synchronization and transient stability in power networks and non-uniform Kuramoto oscillators. In: Proceedings of the American Control Conference. 2010. pp. 930-937

[28] Dörfler F, Chertkov M, Bullo F. Synchronization in complex oscillator networks and smart grids. Proceedings of the National Academy of Sciences of the United States of America. 2013;**110**(6):2005-2010

[29] Simpson-Porco J, Dörfler F, Bullo F. Droop-controlled inverters are Kuramoto oscillators. In: Proceedings of 2012 3rd IFAC Workshop Distributed Estimation and Control in Networked Systems. Vol. 45, No. 26. 2012. pp. 264-269

[30] Simpson-Porco JW, Dörfler F, Bullo F. Synchronization and power sharing for droop-controlled inverters in islanded microgrids. Automatica. 2013;**49**(9):2603-2611

[31] Johnson BB, Sinha M, Ainsworth NG, Dörfler F, Dhople SV. Synthesizing virtual oscillators to control islanded inverters. IEEE Transactions on Power Electronics. 2016;**31**(8):6002-6015

[32] Dörfler F, Bullo F. On the critical coupling for Kuramoto oscillators. SIAM Journal on Applied Dynamical Systems. 2011;**10**(3):1070-1099

[33] Jadbabaie A, Motee N, Barahona M. On the stability of the Kuramoto model of coupled nonlinear oscillators. In:

Proceedings of the American Control Conference. 2004. pp. 4296-4301

[34] van Hemmen JL, Wreszinski WF. Lyapunov function for the Kuramoto model of nonlinearly coupled oscillators. Journal of Statistical Physics. 1993;**72**(1): 145-166

[35] Mirollo RE, Strogatz SH. The spectrum of the locked state for the Kuramoto model of coupled oscillators. Physica D: Nonlinear Phenomena. 2005; **205**(1–4):249-266

[36] Canale E, Monzon P. Almost global synchronization of symmetric Kuramoto coupled oscillators. In: Systems Structure and Control. London, UK: InTech Education and Publishing; 2008. pp. 167-190

[37] Ainsworth N, Grijalva S. A structure-preserving model and sufficient condition for frequency synchronization of lossless droop inverter-based ac networks. IEEE Transactions on Power Apparatus and Systems. 2013;**28**: 4310-4319

[38] Fattahi J, Schriemer H. Algebraic connectivity conditions for synchronization in low-inertia microgrids with adaptive droop-controlled inverters. International Journal of Electrical Power & Energy Systems. 2019;**111**:493-507

[39] Mao Y, Zhang Z. Distributed frequency synchronization and phase-difference tracking for kuramoto oscillators and its application to islanded microgrids. In: IEEE 55th conference on decision and control (CDC), Las Vegas, NV, USA. 2016. pp. 4364-4369

[40] Zhu L, Hill D. Transient stability analysis of microgrids with network-preserving structure. In: 6th IFAC Workshop on Distributed Estimation and Control in Networked Systems. Vol. 49. 2016. pp. 339-344

[41] Zhu L, Hill DJ. Synchronization of power systems and Kuramoto oscillators: A regional stability framework. IEEE Transactions on Automatic Control. 2018. pp. 1-3

[42] Smith D, Jafarpour S, Bullo F. Transient stability of droop-controlled inverter networks with operating constraints. arXiv:1907.05532 [eess.SY]. 2019

[43] Yan L, Chen W, Li C, Dai H, Zhu X, Gan Z. Consensus-based distributed power control in power grids. Science China Information Sciences. 2020;**63**

[44] Rohden M, Sorge A, Timme M, Witthaut D. Self-organized synchronization in decentralized power grids. Physical Review Letters. Aug 2012; **109**(6):064101

[45] Giraldo J, Mojica-Nava E, Quijano N. Synchronization of dynamical networks with a communication infrastructure: A smart grid application. In: 52nd IEEE Conf. Decision and Control. Los Angeles, CA, USA; 2013. pp. 4638-4643

[46] Giraldo J, Mojica E, Quijano N. Synchronization of Heterogeneous Kuramoto Oscillators with Sampled Information and a Constant Leader. International Journal of Control. 2019; **92**:2591-2607

[47] Bergen AR, Hill DJ. A structure preserving model for power system stability analysis. Power Apparatus and Systems, IEEE Transactions on. 1981; **100**(1):25-35

[48] Pecora LM, Carroll TL. Master stability functions for synchronized coupled systems. Physical Review Letters. 1998;**80**:2109-2118

[49] Arenas A, Díaz-Guilera A, Kurths J, Moreno Y, Zhou C. Synchronization in complex networks. Physics Reports. 2008;**469**:93-153

[50] Sun J, Bollt EM, Nishikawa T. Master stability functions for coupled nearly identical dynamical systems. EPL (Europhys Lett). 2009;**85**(6):60011

[51] Motter AE, Myers SA, Anghel M, Nishikawa T. Spontaneous synchrony in power grid networks. Nature Physics. 2013;**9**(3):191-197

[52] Choi YP, Ha SY, Yun SB. Complete synchronization of Kuramoto oscillators with finite inertia. Physica D. 2011; **240**(1):32-44

[53] Ha SY, Kim Y, Li Z. Large-Time Dynamics of Kuramoto Oscillators under the Effects of Inertia and Frustration. SIAM Journal on Applied Dynamical Systems. 2014;**13**(1):466-492

[54] Choi Y-P, Li Z, Ha S-Y, Xue X, Yun S-B. Complete entrainment of Kuramoto oscillators with inertia on networks via gradient-like flow. Journal of Differential Equations. 2014;**257**(7): 2591-2621

[55] Choi Y-P, Li Z. Synchronization of nonuniform Kuramoto oscillators for power grids with general connectivity and dampings. Nonlinearity. 2019;**32**:559

[56] Zhao C, Topcu U, Li N, Low S. Design and stability of load-side primary frequency control in power systems. Automatic Control, IEEE Transactions on. 2014;**59**(5):1177-1189

[57] Olfati-Saber R. Flocking for multi-agent dynamic systems: Algorithms and theory. Automatic Control, IEEE Transactions on. 2006;**51**(3):401-420

[58] Ren W, Beard RW, Atkins E. Information consensus in multivehicle cooperative control: Collective group behavior through local interaction. Control Systems Magazine, IEEE. 2007; **27**(2):71-82

[59] Dimarogonas DV, Johansson KH. Stability analysis for multiagent systems using the incidence matrix: Quantized communication and formation control. Automatica. 2010;**46**:695-700

[60] Pai MA, Padiyar KR, RadhaKrishna C. Transient stability analysis of multi-machine AC/DC power systems via energy-function method. Power Engineering Review, IEEE. 1981; **12**:49-50

[61] Chang H-D, Chu C-C, Cauley G. Direct stability analysis of electric power systems using energy functions: theory, applications, and perspective. Proceedings of the IEEE. 1995;**83**(11): 1497-1529

[62] Chiang H-D, Thorp JS. The closest unstable equilibrium point method for power system dynamic security assessment. Circuits and Systems, IEEE Transactions on. 1989;**36**(9):1187-1200

[63] Liu C-W, Thorp JS. A novel method to compute the closest unstable equilibrium point for transient stability region estimate in power systems. Circuits and Systems I: Fundamental Theory and Applications, IEEE Transactions on. 1997;**44**(7):630-635

[64] Chen L, Min Y, Xu F, Wang K-P. A continuation-based method to compute the relevant unstable equilibrium points for power system transient stability analysis. Power Systems, IEEE Transactions on. 2009;**24**(1):165-172

[65] Chiang H-D, Wu FF, Varaiya PP. A BCU method for direct analysis of power

system transient stability. Power Systems, IEEE Transactions on. 1994; **9**(3):1194-1208

[66] Odor G, Hartmann B. Heterogeneity effects in power grid network models. Physical Review E. 2018;**98**:022305

[67] Grzybowsk JMV, Macau EEN, Yoneyama T. On synchronization in power-grids modelled as networks of second-order Kuramoto oscillators. Chaos. 2016;**26**(11):113113(1-12)

[68] Li Z, Xue X, Yu D. Synchronization and transient stability in power grids based on Lojasiewica inequalities. SIAM Journal on Control and Optimization. 2014;**52**(4):2482-2511

[69] Yang L, Guo Y, Chen N, Qian M, Xue X, Yu D. Power System Transient Stability Analysis via Second-Order Non-Uniform Kuramoto Model. Advances in Materials Research. 2014; **19**:1054-1057

[70] Li H, Chen G, Liao X, Huang T. "Attraction Region Seeking for Power Grids" IEEE Trans. Circuits Syst. II Express Briefs. 2017;**64**:201-205

[71] Zhu L, Hill DJ. Stability Analysis of Power Systems: A Network Synchronization Perspective. SIAM Journal on Control and Optimization. 2018;**56**:1640-1664

[72] Hill DJ, Bergen AR. Stability analysis of multimachine power networks with linear frequency dependent loads. IEEE Trans. Circuits Syst. 1982;**29**: 840-848

[73] Guo Y, Wang Q, Yang L, Zhang D, Xue X, Daren Y. A transient stability analysis method based on second-order nonuniform Kuramoto model. International Transaction on Electrical Energy Systems. 2020;**30**(3):12241(1-17)

[74] Coh-1, Susuki Y, Mezić I, Hikihara T. Coherent Swing Instability of Power Grids. Journal of Nonlinear Science. 2011;**21**(3):403-439

[75] Omelchenko OE. Coherence–incoherence patterns in a ring of non-locally coupled phase oscillators. Nonlinearity. 2013;**26**(9):2469-2498

[76] Wolfrum M, Omelchenko OE, Yanchuk S, Maistrenko YL. Spectral properties of chimera states. Chaos. Mar 2011;**21**(1):013112

[77] Ashwin P, Burylko O. Weak chimeras in minimal networks of coupled phase oscillators. Chaos. Jan 2015;**25**(1):013106

[78] Panaggio MJ, Abrams DM, Ashwin P, Laing CR. Chimera states in networks of phase oscillators: The case of two small populations. Physical Review Letters. Jan 2016;**93**(1): 012218

[79] Belykh IV, Brister BN, Belykh VN. Bistability of patterns of synchrony in Kuramoto oscillators with inertia. Chaos. Sep 2016;**26**(9):094822(1-11)

[80] Li B, Wong KYM. Optimizing Synchronization Stability of the Kuramoto Model in Complex Networks and Power Grids. Physical Review E. 2017;**95**:012207

[81] Ji P. Synchronization in the Second-order Kuramoto Model, Discontinuous phase transitions with frequency-degree correlations and basin stability. Ph.D. Dissertation,. Faculty of Mathematics and Natural Sciences, The Humboldt University of Berlin; 2015

[82] Farhangi R, Beheshti MT. Transient stability conditions assessment in smart grids via new Lyapunov approaches. In:

10th Smart Grid Conference (SGC). Kashan, Iran; 2020. pp. 1-6. DOI: 10.1109/SGC52076.2020.9335744

[83] Farhangi R, Beheshti MT, Jamil M, Khan AA. An analytical approach for the stability analysis of power networks through Kuramoto oscillators model. Communications in Nonlinear Science and Numerical Simulation. 2023;**126**: 107467. DOI: 10.1016/j.cnsns.2023.107467

[84] Lin Z, Francis B, Maggiore M. State agreement for continuous-time coupled nonlinear systems. SIAM Journal on Control and Optimization. 2007;**46**(1): 288-307

[85] Naser MFM. Nonsmooth Lyapunov Stability of Differential Equations. Applied Mathematical Sciences. 2017; **11**(18):887-897

[86] Bacciotti A, Ceragioli F. Stability and Stabilization of Discontinuous Systems and Nonsmooth Lyapunov Functions. ESAIM: Control, Optimization and Calculus of Variations. 2002:361-376

[87] Clarke FH, Ledyaev YS, Stern RJ, Wolenski RR. Nonsmooth Analysis and Control Theory. Springer; 1899

[88] Camlibel MK, Pang J-S, Shen J. Lyapunov stability of complementarity and extended systems. SIAM Journal on Optimization. Jan 2006;**17**(4):1056-1101

[89] Shevitz D, Paden B. Lyapunov stability theory of nonsmooth systems. IEEE Transactions on Automatic Control. Sep 1994;**39**(9):1910-1914

[90] Rantzer A, Ruffer BS, Dirr G. Separable Lyapunov functions for monotone systems. In: IEEE Conference on Decision and Control, Firenze, Italy. 2013. pp. 4590-4594

[91] Dirr G, Ito H, Rantzer A, Ruffer BS. Separable Lyapunov functions for monotone systems: Constructions and limitations. Discrete and Continuous Dynamical Systems - Series B. 2015: 2497-2526

[92] Ito H, Rüffer BS, Rantzer A. Max- and sum-separable Lyapunov functions for monotone systems and their level sets. In: 53rd IEEE Conference on Decision and Control. 2014;**46**(22): 339-344

[93] Moreau L. Stability of continuous-time distributed consensus algorithms. IEEE Conference on Decision and Control. 2004:3998-4003

[94] Clarke FH. Generalized gradients and applications. Transactions of the American Mathematical Society. 1975: 247-262

[95] Danskin JM. The theory of max-min, with applications. SIAM Journal on Applied Mathematics. 1966;**14**(4): 641-664

[96] Farhangi R, Beheshti MTH. The Kuramoto model: The stability conditions in the presence of phase shift. Neural Processing Letters. 2021;**53**: 2631-2648

Chapter 6

Adoptive Inverter Controller for Microgrid in Islanded Mode

Bangar Raju Lingampalli, Subba Rao Kotamraju, Kiran Kumar Malligunta, Bishnu Prasad Muni and Rahul S. Somalwar

Abstract

The utilization of distributed generation (DG) in Microgrids has posed challenges in modeling and operation and has been resolved with power electronic-based interfacing inverters and associated controllers. The inverter controller in both transient and steady states is of paramount importance, as the stability of Microgrid in grid-connected or islanded mode is dependent on inverter control. Again, the type of the network whether low-voltage or high-voltage distribution sector is of prime importance. In high-voltage network, the reactance/resistance ratio is high and hence no issue of decoupling. But in low-voltage networks as this ratio is low and decoupling is an issue. Normally, the Microgrid is located nearer to load centers. Microgrids are active networks with bi-directional current flow. This research work consists of modeling the Microgrid in a transient state, with a special focus on the study of dynamic and static loads pattern in islanded mode. This chapter presents adoptive control of inverter in an islanded Microgrid. A special focus is given to the state space modeling of all the components of the network and then developing linearized models. The sinusoidal time-varying quantities have been transformed into park transformation, to mitigate decoupling.

Keywords: distributed energy resources, X/R ratio, renewable energy sources, dq0, microgrid islanded mode, dynamic modeling, state space, transient state, distributed generation

1. Introduction

The latest civilized nation needs reliable and quality power for day-to-day domestic, commercial, transport, entertainment, etc. The utilization of Renewable Energy Sources (RESs) in Microgrids is not only reducing the Green House Gases (GHGs) but also adhering to the latest regulations of IEEE-1547-2018. Then, the passive distribution networks are becoming active networks with the bidirectional inverters. The distribution generators are small powered renewable energies from solar, wind, micro-turbines, fuel cells, battery systems, compressed air, super capacitors, which are connected in the low-voltage network, nearer to the loads. As the number of small resources is of smaller capacity, the dynamic modeling is a bit complex due to more variable parameters. To

IntechOpen

attain optimal power flow, proportional load sharing of inverter-based micro-sources and to take care of the transient states for voltage and frequency regulation, new control strategy controllers are to be designed, for Microgrid in islanded mode. The chapter comprises of six sections. The literature review is done in Section 2 with a comparison to the proposed methodology in this chapter. The mathematical equations and dynamic modeling are discussed in Section 3. Section 4 depicts network and MATLAB design parameters. Section 5 gives out the MATLAB results and discussion. Section 6 finally gives out conclusions and future work proposals.

2. Literature review

The references of about 10 mentioned in the references section have been reviewed in this section.

Juan C. Vasquez et al. [1]: proposed an adoptive droop controller for voltage source inverter, which can cater to the grid and islanded mode. The controller monitors the parameters of voltage, frequency, and virtual impedance amplitude and angle. The controller dynamics are controlled by decoupling to control the power flow in both grid and islanded modes. The system stability is also assessed by plotting the poles and zeros of the characteristic equation in real and imaginary axes with root locus method. However, the system dynamics are better controlled by the proposed method in this chapter, which is validated in MATLAB/Simulink and discussed.

Liang Zhang et al. [2]: the authors proposed a virtual synchronous generator (VSG) technology to address the issues of inertial support in AC/DC Microgrids. But while improving the stability of the system, fast performance is neglected. To accelerate the frequency differential and to reduce the fluctuation in DC voltage, the virtual inertia is dynamically adjusted. The proposed method in this chapter improves the stability of the controller better than the method proposed by authors Liang Zhang et al., which is proved through MATLAB simulations.

Zhiyong Chen et al. [3]: the authors proposed an adoptive sliding mode controller (ASMC), to obviate the issues of performance of PID controllers in Microgrid islanded mode. This method also enhances the system performance by rejection of disturbance. This method depends upon the nonlinear model of the controller for the performance of interfacing inverters of islanded Microgrid. However, the proposed method in this chapter is capable of managing the dynamics of the system, which is shown in the simulation results.

Qusay Salem et al. [4]: the authors presented a control method with a synchronous reference frame (SRF) with PI controllers for voltage source inverters of DGs. This modified droop control has enhanced Microgrid performance in a grid as well as islanded mode. The proposed method in this chapter can reduce the dynamics and improve the stability, which is shown in MATLAB simulations for different types of loads.

Masoud Dashtdar et al. [5]: the authors presented a self-tuning PI controller with a combination of Genetic Algorithm (GA) and Artificial Neural Network (ANN) to control the frequency of Microgrid in islanded mode. ANN parameters like frequency, voltage, etc., are optimized and adjusted by training with the help of GA, to optimize and adjust the coefficients of the PI controller. The frequency control is located in the secondary control loop of Microgrid. However, the methodology in this chapter is a more robust controller, which will manage the dynamics and help in enhancing the stability of Microgrid. The discussions of simulations in MATLAB platform confirm the proposed methodology in this chapter.

Bahador Fani et al. [6]: the authors have discussed different control topologies of interfaced inverters in islanded mode of Microgrid like droop control techniques, control based on various philosophies like distributed, decentralized and centralized controls, virtual impedance-based control, small signal stability control, etc. The authors suggested future work on the management of inverter fault currents, controller design techniques to control frequency, and voltage while regulating power flow. The proposed method in this chapter is more efficient in getting the reliability of the Microgrid during islanded mode. The method proved its efficiency through the MATLAB simulations.

Bernhard Hammer et al. [7]: the authors developed a method of reducing the oscillations in the Microgrid by tuning controller parameters. The authors have not considered the system speed restriction with the cut-off frequency of the low-pass filter. The proportional load sharing is achieved by tuning the system parameters and balancing the system's natural damping. For this purpose, the authors considered three cascaded controllers. They are to regulate active and reactive powers, to control voltage and the last one is to control current. In this method, the pole assignment method is used, to tune the parameters of the controllers, which may be a bit difficult. However, the method proposed in this chapter is superior to this method, as it can reduce the dynamics and enhance the stability of the system.

Seyed Mohammad Sadegh Hosseinimoghadam et al. [8]: the authors proposed a method to overcome the droop control in the islanded Microgrid. The droop control depends on the proportionality of active power to voltage and reactive power to frequency in islanded Microgrid. The proportional load sharing is attained by incorporating a virtual impedance between the inverter and PCC, to overcome the resistive behavior of low-voltage Microgrids. But this method is not suitable in islanded mode of Microgrid. However, the method proposed in this chapter has more flexibility in managing the system dynamics during transient states.

Michele Tucci et al. [9]: the authors suggested a synthetic control method for stabilizing frequency and voltage control of the Microgrid in the islanded mode with DGs including power lines and loads. This method is suitable to use with plug and play resources also the modeling of power lines and DGs are not required. The regulator tuning is not required when new DGs are plugged into islanded Microgrid. The model cannot regulate power flow, as it cannot work in parallel with local regulators. At this juncture, the method proposed in this chapter can work efficiently as it is having cascaded controller, which is proved with simulations in MATLAB.

Lasantha Meegahapola et al. [10]: the authors proposed a strategy for Microgrid operation and control for seamless synchronous islanding from grid mode by keeping the dynamics of frequency and phase angle, without interruption to sensitive loads. The method is so efficient while resynchronizing to the grid without disrupting the system dynamics. The synchro-phasor problems and telecom delays are hurdles in maintaining the system stability. But this chapter's methodology is more superior in managing the system and the low inertia effect. The MATLAB simulations proved the effectiveness of this method over the referred author's strategy.

3. Modeling of islanded microgrid with loads

The Microgrid enters an islanded mode in case of non-availability of the grid or when grid parameters are beyond tolerance limits. The important factors to be considered in modeling islanded Microgrid are the optimum values for sizing, placing, accuracy, and validity in different modules [11]. These can be like generation module, network module, and load module. The model may be linear dynamic, utilizing a synchronously rotating reference frame or stationary frame [12].

The load module can be sub-divided into active, passive, and dynamic load modules. The dynamic modules may be rectifier interfaced load (RIAL) or induction motor load (IM). The modeling of different modules is done in the following sections.

3.1 Generation controller

The controller of the distributed generation inverter comprises of three loops. The first loop is the outer power loop which adjusts the voltage amplitude of the inverter output according to the droop control philosophy. The second and third loops are the voltage loop and current loop, which regulate the frequency variations and create damping for the output filter. **Figure 1** shows the details of the generation controller.

Transforming the mathematical model into state variables, the virtual control outputs are taken to power, voltage and current controllers, to get the references [6]. The low-pass filter filters the instantaneous power components passing through it. The selection of low-pass filter bandwidth is so selected that the coefficients of droop control will have the control on damping and oscillations of the inverter, so that the control and stability are achieved [13]. This phenomenon will reduce the circulating currents, and thus, the damage due to the overloading of renewable sources is mitigated [14]. The islanded inverter interfaced DG reference voltage and frequency are given by,

$$\omega_{dr}^{*} = \omega_n - m_p \, \mathrm{P}. \tag{1}$$

$$v_{od-dr}^{*} = V_n - n_Q \, \mathrm{Q}. \tag{2}$$

where ω_n and V_n are nominal set values of d-axis frequency and voltage as, v_{oq-dr}^{*} is set to zero, to align the q-axis voltage to x-axis on cartesian co-ordinates and hence the reactive power and active power droops for the standard range are given by,

$$m_p = \omega_{max} - \omega_{min} / P_{max} \tag{3}$$

$$n_Q = v_{od_{max}} - v_{od_{min}} / Q_{max} \tag{4}$$

The average values of reactive and active powers through the LCL filter in **Figure 2** are given by,

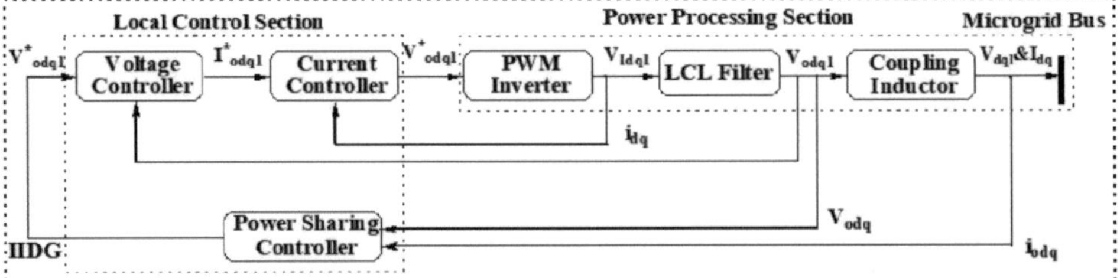

Figure 1.
Power and control diagram of inverter interfaced DG.

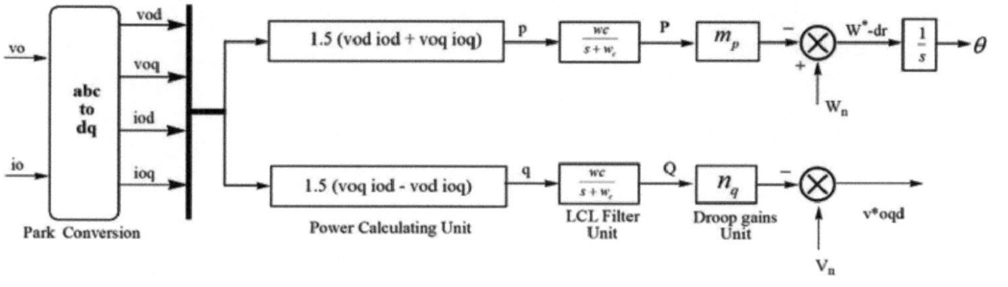

Figure 2.
Outer loop power controller.

$$\mathrm{P} = \frac{\omega_c}{s+\omega_c} \times 1.5\left(v_{od}\, i_{od} + v_{oq} i_{oq}\right) \tag{5}$$

$$\mathrm{Q} = \frac{\omega_c}{s+\omega_c} \times 1.5\left(v_{oq}\, i_{od} - v_{od}\, i_{oq}\right). \tag{6}$$

where 1.5 $(v_{od}\, i_{od} + v_{oq} i_{oq})$ is active power and 1.5 $(v_{od}\, i_{od} + v_{oq} i_{oq})$ is reactive power. The parameters $v_{od}\, i_{od}$ and $v_{oq} i_{oq}$ are output voltage and current of "dq" reference frame and the LCL filter cutoff frequency is ω_c.

3.1.1 Controllers of inner loop voltage and current control

The aim of the current and voltage regulators is to mitigate the high-frequency disturbances and to provide natural damping to the filter circuit [15]. The reference current to the filter inductor i_{ldq} is given by voltage controller and v_{idq} is given by current controller are to be given to PWM module as shown in **Figure 3**. The voltage controller regulates the voltage and the current controller modifies the voltage wave of the filter inductor [16]. PI controllers are used in voltage and current controllers with feed forward and backward terms. Both these controllers control the output voltage and filter inductor current [17]. The proportional and integral gains of these PI controllers along with forward gain are given by,

K_{pv}—voltage controller proportional gain, K_{iv}—voltage controller integral gain, K_{pi}—current controller proportional gain, K_{ii}—current controller integral gain, and

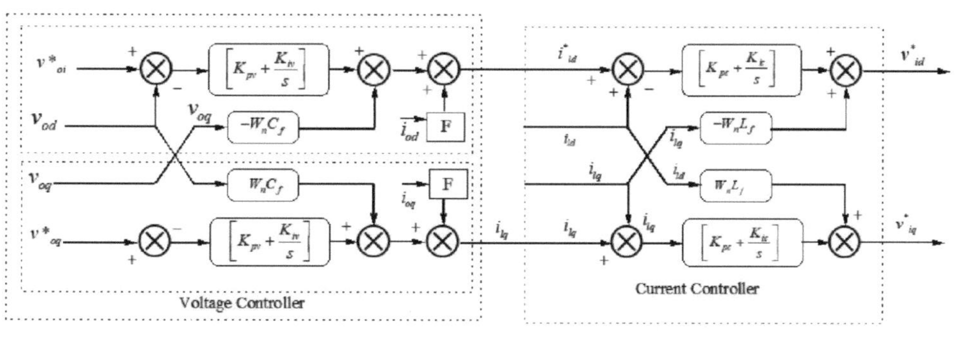

Figure 3.
Inverter controllers of DG for voltage and current.

F—feed-forward gain. The dynamical equations of voltage and current controllers are given by,

$$i_{ld}^{*} = K_{pv}\left(v_{od}^{*} - v_{od}\right) + K_{iv}\,\phi_{d} - \omega_{n}\,C_{f}\,v_{oq} + \mathrm{F}\,i_{od} \tag{7}$$

$$i_{lq}^{*} = K_{pv}\left(v_{oq}^{*} - v_{oq}\right) + K_{iv}\,\phi_{q} + \omega_{n}\,C_{f}\,v_{od} + \mathrm{F}\,i_{oq} \tag{8}$$

$$v_{ld}^{*} = K_{pi}\left(i_{ld}^{*} - i_{ld}\right) + K_{ii}\,\lambda_{d} - \omega_{n}\,L_{f}\,i_{lq} \tag{9}$$

$$v_{lq}^{*} = K_{pi}\left(i_{lq}^{*} - i_{lq}\right) + K_{ii}\,\lambda_{q} + \omega_{n}\,L_{f}\,i_{ld} \tag{10}$$

In which voltage controller reference currents are i_{ld}^{*} and i_{lq}^{*}, and used as set values to the current controller. The current controller reference voltage magnitude and angles are given by v_{ld}^{*}, v_{lq}^{*} and λ_d, λ_q respectively, and are used as reference set values to PWM—Pulse Width Modulated inverter, which is depicted in **Figure 1**.

3.1.2 Module of LCL filter output and inductance

The LCL filter output along with the inductance of coupling are considered as reference values. Coupling inductance adjusts the inverter output impedance and so decouples the active and reactive powers [18]. The LCL filter (low-pass filter consisting of inductance, capacitance, and inductance) output dynamical voltage and current are given in the following equations [19]. The inter-facing inverter generates voltage which is equal to the reference voltage, $v_i = v_i^*$, as depicted in **Figure 3**. The d- and q-axes reference currents and voltages are given by,

$$i_{od}^{*} = \frac{1}{L_c}\left(v_{od} - v_{md} - r_c\,i_{od}\right) + \omega\,i_{oq} \tag{11}$$

$$i_{oq}^{*} = \frac{1}{L_c}\left(v_{oq} - v_{mq} - r_c\,i_{oq}\right) - \omega\,i_{od} \tag{12}$$

$$i_{ld}^{*} = \frac{1}{L_f}\left(v_{ld} - v_{od} - r_f\,i_{ld}\right) + \omega\,i_{lq} \tag{13}$$

$$i_{lq}^{*} = \frac{1}{L_{ff}}\left(v_{lq} - v_{oq} - r_f\,i_{lq}\right) + \omega\,i_{ld} \tag{14}$$

$$v_{od}^{*} = \frac{1}{C_f}\left(i_{ld} - i_{od}\right) + \omega\,v_{oq} + R_d\left(i_{ld}^{*} - i_{od}^{*}\right). \tag{15}$$

$$v_{oq}^{*} = \frac{1}{C_f}\left(i_{lq} - i_{oq}\right) - \omega\,v_{od} + R_d\left(i_{lq}^{*} - i_{oq}^{*}\right). \tag{16}$$

In which v_{md} = Microgrid d-axis voltage, v_{mq} = Microgrid q-axis voltage, r_f, r_c = parasitic resistances, L_f, L_c = parasitic inductances, R_d = damping resistor connected in series with C_f.

3.2 Network control module

In inverter interfaced DGs, the VSIs (Voltage source inverters), the time constant is small and hence the network dynamics have an impact on stability [20]. The network RL form is shown in **Figure 4**. The dynamical equations of the line connected between Microgrid bus and main grid in DQ reference frame (the capital letters DQ shows for network) are given by,

$$i^*_{lineD} = \frac{-R_{line}}{L_{line}} i_{lineD} + \frac{1}{L_{line}} (v_{mD} - v_{gD}) + \omega\, i_{lineQ} \tag{17}$$

$$i^*_{lineQ} = \frac{-R_{line}}{L_{line}} i_{lineQ} + \frac{1}{L_{line}} (v_{mQ} - v_{gQ}) + \omega\, i_{lineD} \tag{18}$$

Eqs. (17) and (18) can be written in the small scale linearized state space,

$$\Delta\, i^*_{lineD} = \frac{-R_{line}}{L_{line}} \Delta i_{lineD} + \frac{1}{L_{line}} (\Delta v_{mD} - \Delta v_{gD}) + \Delta\, \omega_{com}\, I_{lineQ} \tag{19}$$

$$\Delta\, i^*_{lineQ} = \frac{-R_{line}}{L_{line}} \Delta i_{lineQ} + \frac{1}{L_{line}} (\Delta v_{mQ} - \Delta v_{gQ}) + \Delta\, \omega_{com}\, I_{lineD} \tag{20}$$

where $v_{mD}, v_{gD}, v_{mQ}, v_{gQ}$ are the voltages of Microgrid bus and main grid or distribution grid bus in DQ axes, respectively. The $\Delta\, i^*_{lineD}$ and $\Delta\, i^*_{lineQ}$ are the incremental reference current values in DQ axes.

3.3 Load control module

Various types of loads like passive, active, constant power and dynamic loads are switched on and thrown off at different times at Microgrid bus to find out the stability of the islanded Microgrid [21]. MATLAB/Simulink software package is used for simulating the model and the Microgrid interfaced inverter controller of DG found to be stable with proper power regulation, voltage, and frequency deviations with acceptable deviations as prescribed by IEEE-1847-2018 standards [22]. The network circuit, for the total simulations, is shown in **Figure 5**.

3.3.1 Passive load module

The tested passive loads considered are RL load and constant power load CPL [23]. The resistive loads, which can be considered are incandescent lamps, heaters, etc. The electric motors, solenoids, etc., can be categorized as RL loads [24]. The passive load of constant power load (CPL) considered is instantaneous impedance, which is of positive value [25]. The passive load modeling is done in the following sections [26].

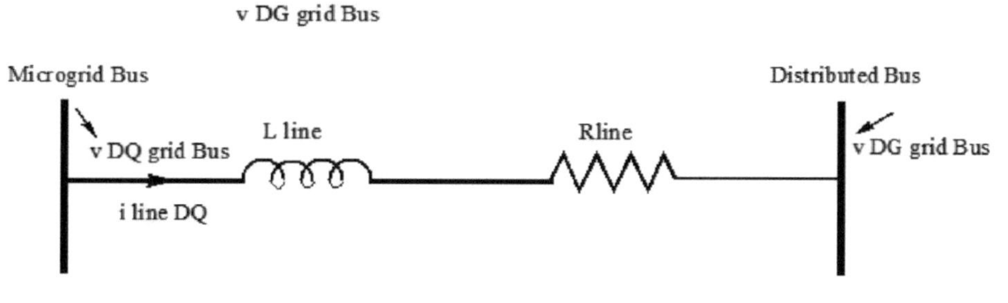

Figure 4.
Line configuration with RL.

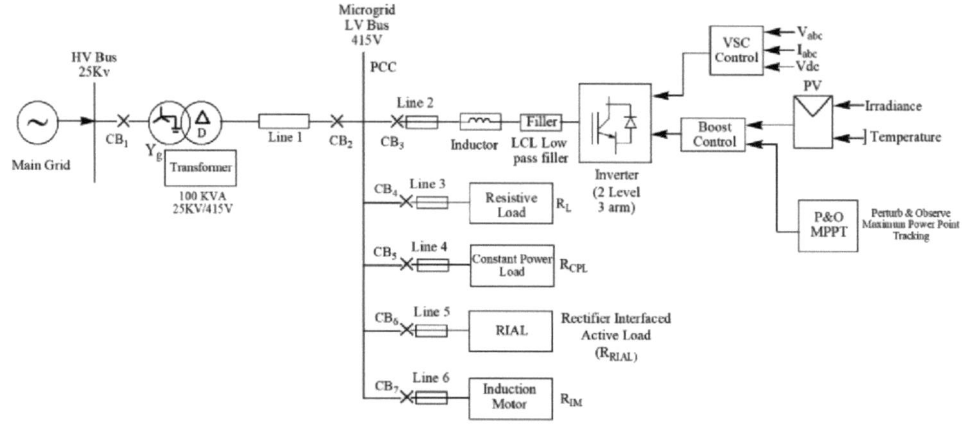

Figure 5.
Total network circuit for testing inverter interfaced DG.

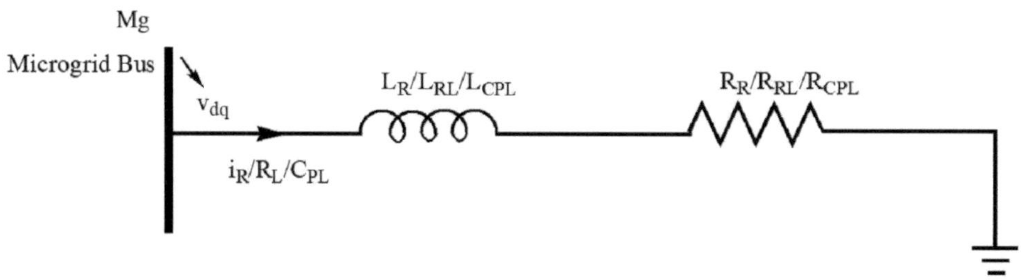

Figure 6.
Passive network model.

The network model for passive loads is shown in **Figure 6**. The passive load dynamical equations, which are connected at PCC, are given by,

$$i^*_{PLD} == \frac{-R_{PL}}{L_{PL}} i_{PLD} + \frac{1}{L_{PL}} (v_{mD}) + \omega\, i_{PLQ} \tag{21}$$

$$i^*_{PLQ} = \frac{-R_{PL}}{L_{PL}} i_{PLQ} + \frac{1}{L_{PL}} (v_{gD}) - \omega\, i_{PLD} \tag{22}$$

Eqs. (21) and (22) are common for passive loads and PL, has to be replaced by R_{LOAD} for resistive, RL_{LOAD} for inductive load and R_{CPL} for constant power load. But constant power load has to be expressed in real and imaginary parts ($R_{CPL} + j\,X_{CPL}$),

For which the real part is $-r_{CPL}\cos\alpha$ and imaginary part is $-r_{CPL}\sin\,\alpha$, where α is angle of delay of rectifier controller.

3.3.2 Module of active load

The UPSs—un-interrupted power systems are nowadays being used in industries and residential buildings. The three-phase inverters are being used extensively in lifts and cranes [27]. The single-line power and control circuit is depicted in **Figure 7**. The control circuit is having of DC voltage control and AC current control part of the Rectifier Interfaced Active Load [28]. These consist of PI controllers. The DC voltage

DOI: http://dx.doi.org/10.5772/intechopen.1003937

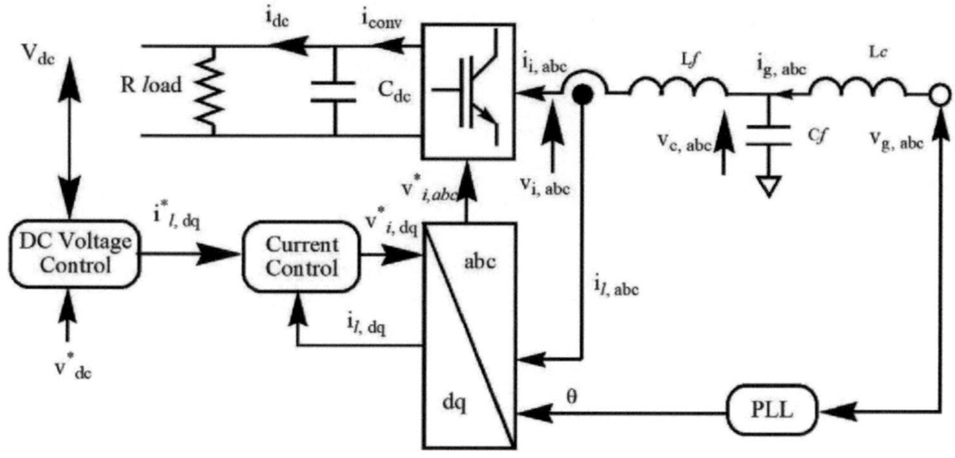

Figure 7.
Control circuit of RIAL – Rectified interfaced active load.

controller controls the voltage of the capacitor C_{dc} and the inner current of L_f is controlled by AC controller [29]. The reference input to AC controllers is given by the DC voltage controller and this in turn controls the current through L_{fAL}. The feed forward terms being given by AC controllers, which helps in decoupling the d and q. The d-q axes inductor currents are given by $i_{ldq\ AL}$, the frequency of nominal system voltage is ω_n and inductance L_{fAL}. The control circuit is shown in **Figure 8**.

3.3.3 Dynamic load module

The dynamic load considered is an induction motor (IM) and the dynamic equations in the D-Q reference frame, considering the resistance (r_S,), inductance ($L_{SS,}$), mutual inductance (L_m) of motor, rotor slip (s), and stator angular frequency (ω) are given by,

$$V_{Qs} = r_s i_{Qs} + L_{ss}\, i^*_{Qs} + L_m\, i^*_{Qr} + \omega\, L_{ss}\, i_{Ds} + \omega\, L_m\, i_{Dr} \tag{23}$$

$$V_{Ds} = r_s i_{Ds} + L_{ss}\, i^*_{Ds} + L_m\, i^*_{Dr} - \omega\, L_{ss}\, i_{Qs} - \omega\, L_m\, i_{Qr} \tag{24}$$

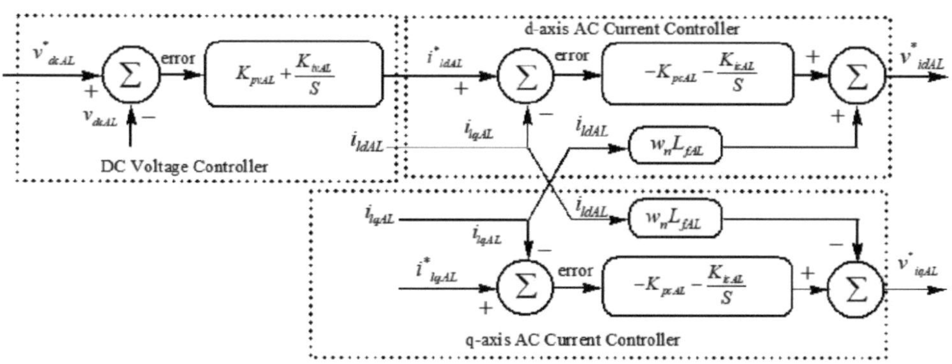

Figure 8.
AC d-q current and DC voltage controllers of RIAL.

$$V_{Qr} = r_S i_{Qr} + L_{rr}\, i^*_{Qr} + L_m\, i^*_{Qs} + s\omega\, L_{rr}\, i_{Dr} + s\omega\, L_m\, i_{Ds} \tag{25}$$

$$V_{Dr} = r_r i_{Dr} + L_{rr}\, i^*_{Dr} + L_m\, i^*_{Ds} - s\omega\, L_{rr}\, i_{Qr} - s\omega\, L_m\, i_{Qs} \tag{26}$$

$$T_e = \frac{3}{2}\frac{P}{2}\, L_m \left(i_{Qs}\, i_{Dr} - i_{Ds} i_{Qr}\right) \tag{27}$$

$$T_e - T_L = \mathrm{J}\, \frac{d}{dt}\, \{(1-s)\omega)\} \tag{28}$$

in which the terms in Eqs. (27) and (28) T_e is motor torque, T_L is load torque, p is the number of poles of the motor and J is the inertia of the combined motor and load. It is assumed that the stator supply frequency is ω having an impact on inverter droop equations [30]. The linearized equations of induction motor considering the small signal deviations are given by,

$$\underbrace{\begin{bmatrix} \Delta_{v_{Qs}} \\ \Delta_{v_{Ds}} \\ \Delta_{v_{Qr}} \\ \Delta_{v_{Dr}} \\ \Delta_{T_L} \end{bmatrix}}_{\Delta_{uIM}} = F_{IM} \underbrace{\begin{bmatrix} \Delta_{i_{Qs}} \\ \Delta_{i_{Ds}} \\ \Delta_{i_{Qr}} \\ \Delta_{i_{Dr}} \\ \Delta_s \end{bmatrix}}_{\Delta_{XIM}} + E_{IM} \underbrace{\begin{bmatrix} \Delta_{i_{Qs}} \\ \Delta_{i_{Ds}} \\ \Delta_{i_{Qr}} \\ \Delta_{i_{Dr}} \\ \Delta_s \end{bmatrix}^*}_{\Delta_{XIM}} + D_{1IM}\Delta\omega + D_{2IM}\Delta\omega^* \tag{29}$$

The linearized small signal in state-space model of the induction motor load is given by,

$$\begin{aligned} \left[\Delta x^*_{IM}\right]_{5\times 1} = {} & [\Delta A_{IM}]_{5\times 5}[\Delta x_{IM}]_{5\times 1} + [B_{1IM}]_{5\times 5}[\Delta u_{IM}]_{5\times 1} \\ & +[B_{2IM}]_{5\times 1}[\Delta\omega]_{1\times 1} + [B_{3IM}]_{5\times 1}[\Delta\omega^*]_{1\times 1} \end{aligned} \tag{30}$$

$$[\Delta y_{IM}]_{5\times 1} = [\Delta C_{IM}]_{5\times 5}\, [\Delta x_{IM}]_{5\times 1} \tag{31}$$

In which A_{IM} = state variable matrix of induction motor load
B_{1IM} = input matrix of voltage
B_{2IM} = input matrix of load torque
B_{3IM} = input matrix of load frequency and derivatives
C_{IM} = output matrix of dynamic load of induction motor
From Eq. (30), Δx_{IM} and Δu_{IM} can be written as,

$$[\Delta x_{IM}] = \begin{bmatrix}\Delta i_{Qs} & \Delta i_{Ds} & \Delta i_{Qr} & \Delta i_{Dr} & \Delta s\end{bmatrix}^T \tag{32}$$

$$[\Delta u_{IM}] = \begin{bmatrix}\Delta v_{Qs} & \Delta v_{Ds} & \Delta v_{Qr} & \Delta v_{Dr} & \Delta T_L\end{bmatrix}^T \tag{33}$$

The small signal linearized equations of induction motor dynamic load are given by,

$$\begin{aligned} [\Delta x^*_{IM}]_{5n_{IM\times 1}} = {} & [\Delta A_{IM}]_{5n_{IM\times}5n_{IM}}[\Delta x_{IM}]_{5n_{IM\times}1} + [B_{1IM}]_{5n_{IM\times}5n_{IM}}[\Delta u_{IM}]_{5n_{IM\times}1} \\ & +[B_{2IM}]_{5n_{IM\times 1}}[\Delta\omega_{IcomM}]_{1_\times 1} + [B_{3IM}]_{5n_{IM\times}1}\,[\Delta\omega^*_{com}]_{1\times 1} \end{aligned} \tag{34}$$

$$[\Delta\quad y_{IM}]_{5n_{IM\times}1} = [\quad C_{IM}]_{5n_{IM\times}5n_{IM}}\,[\Delta\quad x_{IM}]_{5n_{IM\times}1} \tag{35}$$

In which A_{IM} = state matrix of induction motor dynamic load
B_{1IM} = input matrix of voltage

B_{2IM} = input matrix of load torque
B_{3IM} = input matrix of load frequency and derivatives
C_{IM} = output matrix of dynamic load of induction motor
From Eq. (34), Δx_{IM} and Δu_{IM} of linearized transpose matrix form, can be written as,

$$[\Delta x_{IM}] = [\Delta x_{IM1} \quad \Delta x_{IM2} \quad \dots \quad \dots \quad \Delta x_{IMn_{IM}}]^T \tag{36}$$

$$[\Delta u_{IM}] = [\Delta u_{IM1} \quad \Delta u_{IM2} \quad \dots \quad \dots \quad \Delta u_{IMn_{IM}}]^T \tag{37}$$

3.4 Total microgrid model

The total system of the entire state space model in the linearized notation of Microgrid interfaced inverter can be derived from mapping all models of matrixes, which are connected to PCC [31]. The Microgrid may either export power or import from the grid (−ve sign) or import to the grid (+ve sign) depending upon the D, Q matrixes of the network. The generation, net, and load modules are given mapping matrixes [32]. $M_{GEN}(m, g)$ is the DG module with Microgrid bus with "m" and grid bus with "g" of the order $2_{nN} \times 2_{nG}$ with line elements and is given by mapping matrix of generation is given by,

$$M_{GEN}(m, g) = \begin{cases} 1 \text{ if inverter DG is connected to grid at PCC node} \\ 0 \text{ if inverter DG isn't connected to grid at PCC node} \end{cases} \tag{38}$$

The network module mapping matrix with all the connecting lines and nodes is expressed by M_{NET} and is given by,

$$M_{NET}(m, g) = \begin{cases} -1 \text{ if inverter DG is exporting to grid at PCC node} \\ +1 \text{ if inverter DG is importing from grid at PCC node} \end{cases} \tag{39}$$

The mapping matrix of "n" passive and "n" active loads of the order of n_{PL} and n_{AL} is given by,

$$M_{PL}/M_{AL}\ (m, g) = \begin{cases} -1 \text{ if active or passive load is connected to grid at PCC node} \\ 0 \text{ if active or passive load is not connected to grid at PCC node} \end{cases} \tag{40}$$

The dynamic load mapping matrix of induction motor of the order of $n_{nN} \times n_{nIM}$ is given by,

$$M_{IM}(m, g) = \begin{cases} -1 \text{ if IM (induction motor) stator is connected to grid at PCC node} \\ 0 \text{ if IM of stator is not connected to grid at PCC node} \end{cases} \tag{41}$$

The assumption is that all nodal voltages are inputs to every module and currents are outputs [33]. To reduce the deviations in dynamic stability inverter interfaced DG, a large value virtual resistor r_n is introduced in the circuit between every node and ground, which is shown in **Figure 9(a)**. The purpose of inserting this resistance is to reduce the dynamics of the inverter and to improve the stability of the system. The Microgrid bus voltage is given by,

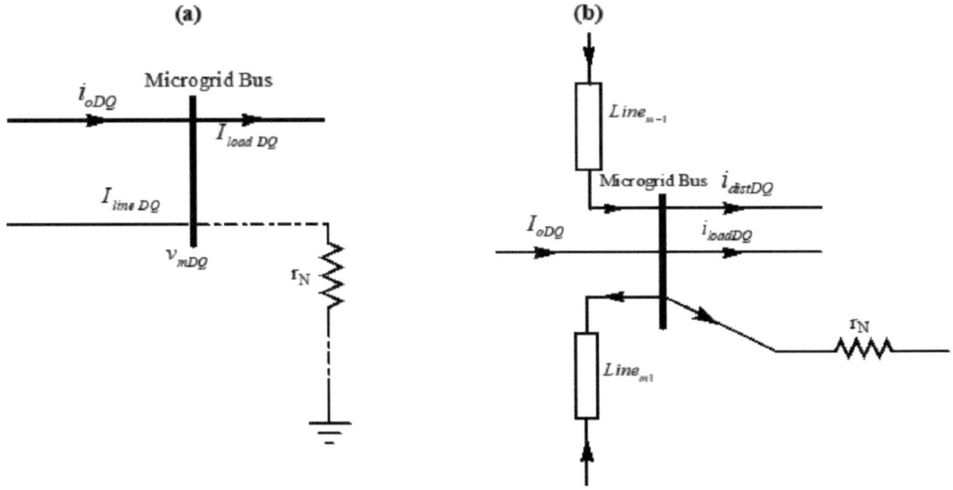

Figure 9.
(a). Virtual resistor r_N *between microgrid bus and ground. (b). Disturbance at grid bus created from microgrid bus.*

$$v_{gDQ} = r_n\left(i_{gDQ} + i_{lineDQ} - i_{lineDQ}\right) \tag{42}$$

To analyze the small signal stability, a disturbance is created at the Microgrid bus by switching on sudden dynamic load and throwing off, and the response is recorded [34]. There are no remarkable deviations in frequency and voltage, which are as per IEEE-1547-2018, it should be in the limits for voltage $0.88 \leq v \leq 1.11$p.u., and for frequency $49.5 \leq Hz \leq 50.5$. **Figure 9(b)** depicts the scenario of load current [35]. The corresponding nodal currents and voltages in terms of mapping matrixes are given by,

$$\begin{aligned}\left[v_{gDQ}\right]_{2n_{N\times}1} = &\left[M_{GEN}\right]_{2n_{N\times}2n_G}\left[\Delta i_{ODQ}\right]_{2n_{G\times}1} + \left[M_{NET}\right]_{2n_{N\times}2n_L}\left[\Delta i_{LINEDQ}\right]_{L1} \\ &+\left[M_{PL}\right]_{2n_{N\times}2n_L}\left[\Delta i_{PLDQ}\right]_{2nPL\times1} + \left[M_{AL}\right]_{2n_{N\times}2n_{AL}}\left[\Delta i_{ALDQ}\right]_{2nAL\times1} \\ &+\left[M_{IM}\right]_{2n_{N\times}2n_{IM}}\left[\Delta i_{DQ}\right]_{2nIM\times1} + \left[R_N\right]_{2n_{N\times}2n_N}\left[M_{dist}\right]_{2n_{N\times}2}\left[\Delta\, i_{distDQ}\right]_{2\times1}\end{aligned} \tag{43}$$

In which M_{dist} is the disturbance mapping matrix of order $2n_{N\times}2$ and maps the load disturbance node into network node. Hence the M_{dist} elements are given by,

$$M_{dist}(m, g) = \begin{cases} -1 \text{ if load disturbance is connected to grid at PCC node} \\ 0 \text{ if load disturbance is not connected to grid at PCC node} \end{cases} \tag{44}$$

4. Design parameters

The design values calculated and used in this work are given in the **Tables 1–3**.

5. Simulation results and discussion

The model which is developed, is validated through simulations in MATLAB/ Simulink platform. There are four types of loads considered here for analysis

Line	Value and units
1	(0.4 + j038) Ω
2	(0.5 + j0.25) Ω
3	(0.23 + j0.11) Ω
4	(0.35 + j0.58) Ω
5	(0.3 + j0.47) Ω
6	(0.25 + j0.15) Ω

Table 1.
Parameters of line.

Component	Value and units
DG and Inverter output	100kVA
Filter Resistance, R_f	0.22 Ω
Filter Inductance, L_f	1.45 mH
Filter Resistance, C_f	5.5 μF
Cut off frequency, ω_c	31.41 rad/sec
proportional gain constant of Voltage controller, K_{pv}	0.08
integral gain constant of Voltage controller, K_{iv}	400
proportional gain constant of Current controller, K_{pi}	11.5
proportional gain constant of Current controller, K_{ii}	14,000
Feed forward gain, F	0.75
Coupling Resistor, R_c	0.02 Ω
Coupling Inductor, L_c	0.4 mH

Table 2.
Inverter interfaced DG parameters.

Load	Parameter	Value and unit
Resistive Load (RL)	Active Power Load (AL)	18 kW
	Reactive power load (Inductive)	10 kVA
Constant Power Load (CPL)	Active Power	20 kW
	Reactive power load (Inductive + Capacitive)	10 kVA + 10 kVAr
	$\cos\alpha$ (α- rectifier delay angle)	0.86
Rectifier Interfaced Active Load (RIAL) with DC Motor	Armature Resistance R_a	0.6 Ω
	Armature Inductance L_a	0.12 H
	Field Resistance R_f	240 Ω
	Field inductance, L_f	120 H
	Field-armature mutual inductance, L_{af}	1.8 H

Load	Parameter	Value and unit
	Armature voltage, V_a	300 V
	Field voltage, V_f	240 V
	Motor rating	5 HP
	Motor speed	1750 rpm
Induction Motor Load (IM)	Motor Capacity	7.5HP/10 kVA
	Frequency	50 Hz
	Speed	1500 rpm
	Volts	415 V
	Number of poles, P	4
	Stator resistance/Leakage inductance	1.77 Ω/0.0139H R_s/Z_s
	Rotor resistance/Leakage inductance	1.34 Ω/0.0121H R_r/Z_r
	Magnetizing inductance	(L_m) 0.3687H
	Torque, T_L	48.83 Nw.m

Table 3.
Load parameters.

purposes, which are passive load RL, constant power load CPL, RIAL-rectifier interfaced active load and induction motor dynamic load. The switching time for all loads is 0.4 secs and the switching effect is analyzed at PCC.

5.1 Passive load (R_L)

The Microgrid model developed is tested for analyzing the effect of sudden transients like loads switching on and throwing off with the passive load, which consists of resistance and inductance with the values of R_L 18 kW and L_L 10 kVA, respectively. The load is switched on at PCC through the circuit breaker at 0.4 secs and thrown off at 0.8 secs, over a sampling time of 1 sec in MATLAB/Simulink. The parameters such as voltage and frequency are monitored for deviations and found to be within limits. The controller is stable without any issue of dynamics and the profiles of frequency, voltage and active and reactive powers are investigated at PCC, which are shown in **Figures 10–13.**

5.2 Passive load (CPL)

The passive load connected at PCC, in this case, is resistance, inductive, and capacitive reactance, with the values as shown in **Table 3**. The load is switched on at PCC through the circuit breaker at 0.4 secs and thrown off at 0.8 secs, over a sampling time of 1 sec in MATLAB/Simulink. The frequency deviations are greater while throwing off at 0.8 secs than at switching on an instant of 0.4 secs, but well within the standards of IEEE. Similarly, the voltage deviations are more at the time of throwing off, but within limits. The controller is stable without any issue of dynamics and the profiles of frequency, voltage, active, and reactive powers are shown in **Figures 14–17.**

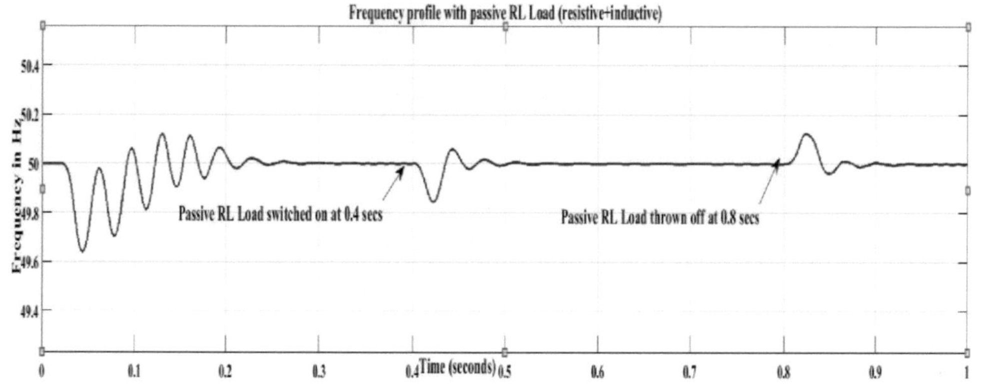

Figure 10.
Frequency profile with R_L load.

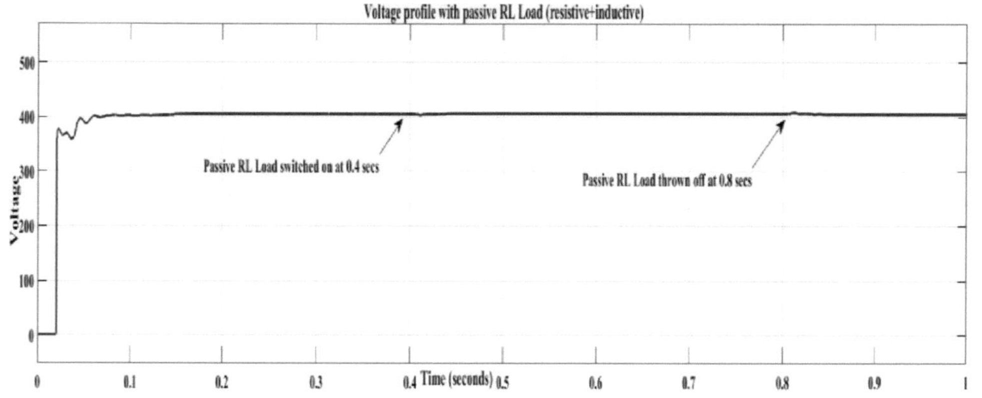

Figure 11.
Voltage profile with R_L load.

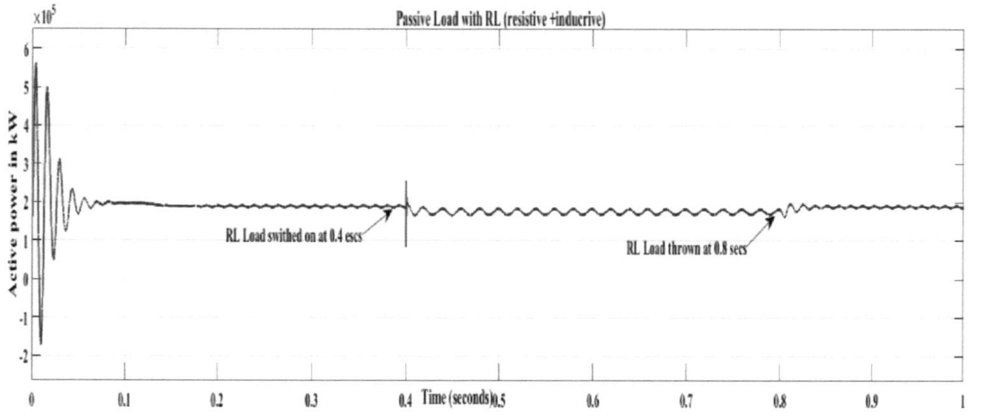

Figure 12.
Active power profile with R_L load.

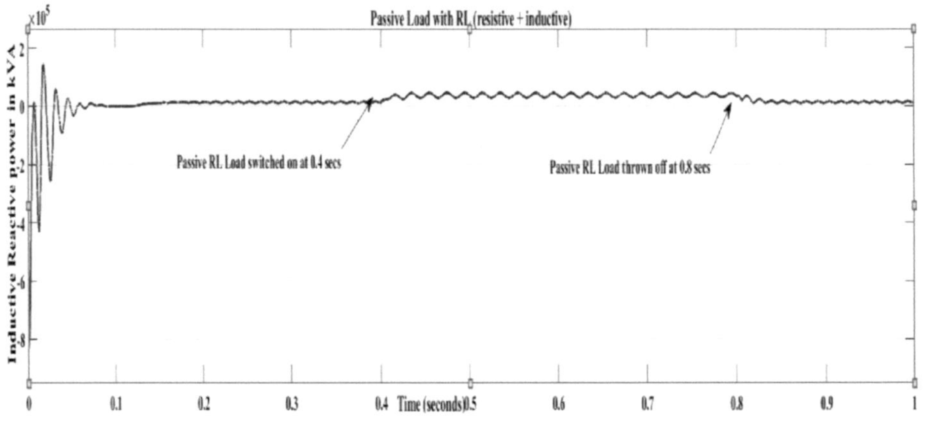

Figure 13.
Reactive power profile with R_L load (inductive reactive power).

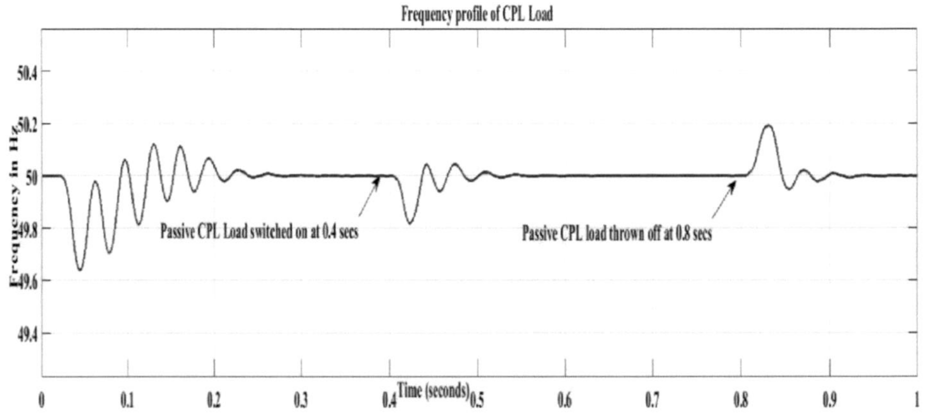

Figure 14.
Frequency profile with CPL load.

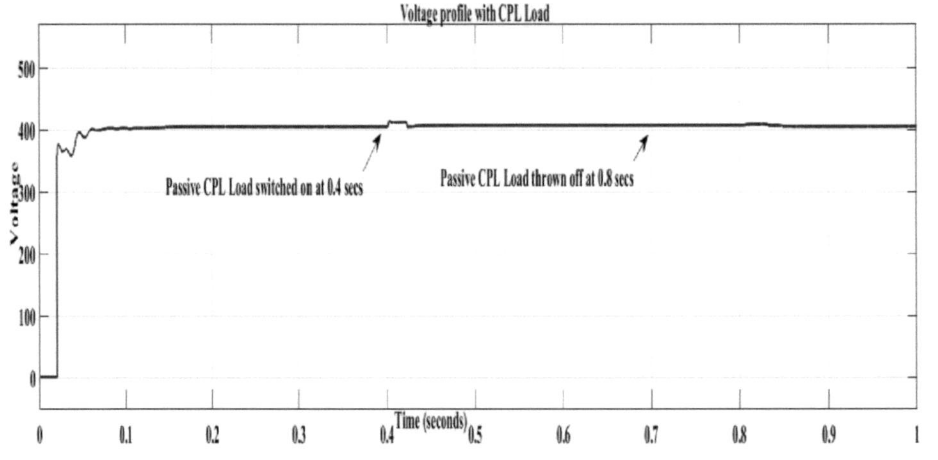

Figure 15.
Voltage profile with CPL load.

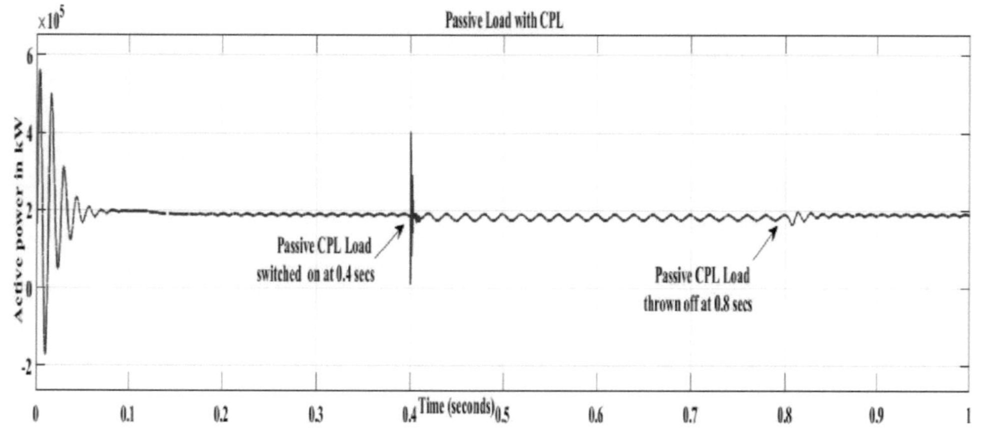

Figure 16.
Active power profile with CPL load.

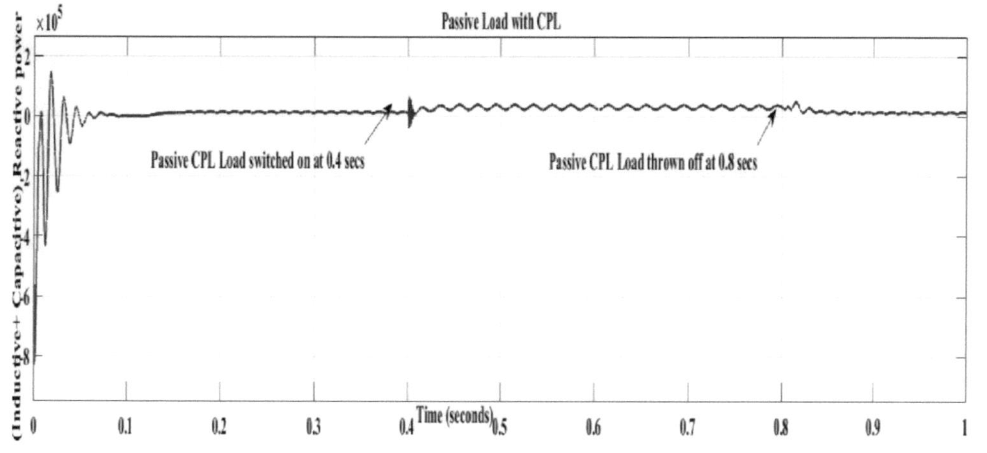

Figure 17.
Reactive power profile with CPL load (resistive, inductive and capacitive).

5.3 Rectifier interfaced active power (RIAL)

The three-phase full-wave bridge rectifier is connected to a DC Motor of 5 HP, 1750 rpm with variable field excitation 240/220/200 volts. At all three volts, the speed, armature current, field current, and torque are obtained in MATLAB/Simulink software and the results are depicted in **Figures 18–20**. As the field voltage is varied from 200 to 240 volts, the speed and torque characteristics are shown in the Figures below. The PCC frequency and voltages are well within permissible limits as per IEEE-1547-2018, 49.5 $\leq f \leq$ 50.5 Hz and voltage 0.88 $\leq V \leq$ 1.11 Volts. This indicates that the adoptive controller performance is as per the design values in the islanded Microgrid.

5.4 Dynamic load with induction motor (IM)

The induction motor dynamic load is connected at PCC with a 10 HP Motor, 1500 rpm. The switching on of the induction motor at PCC with full load, created some transients initially but the controller is stable to absorb these dynamics, which is shown

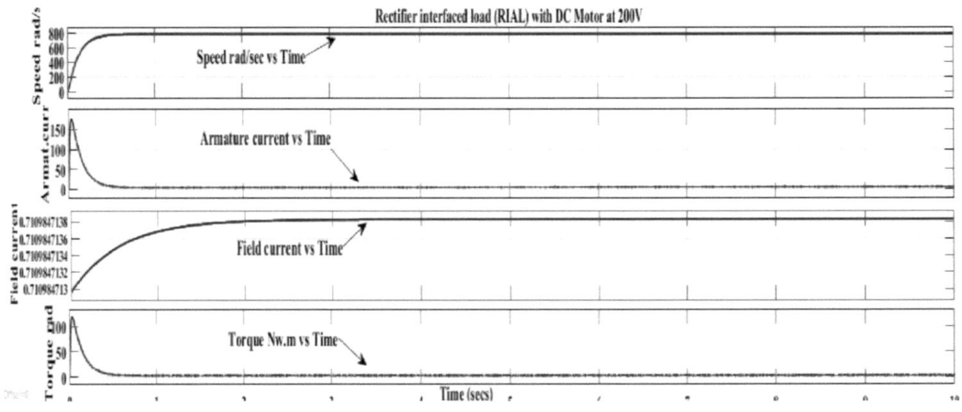

Figure 18.
DC motor characteristic curves with field voltage 200 volts.

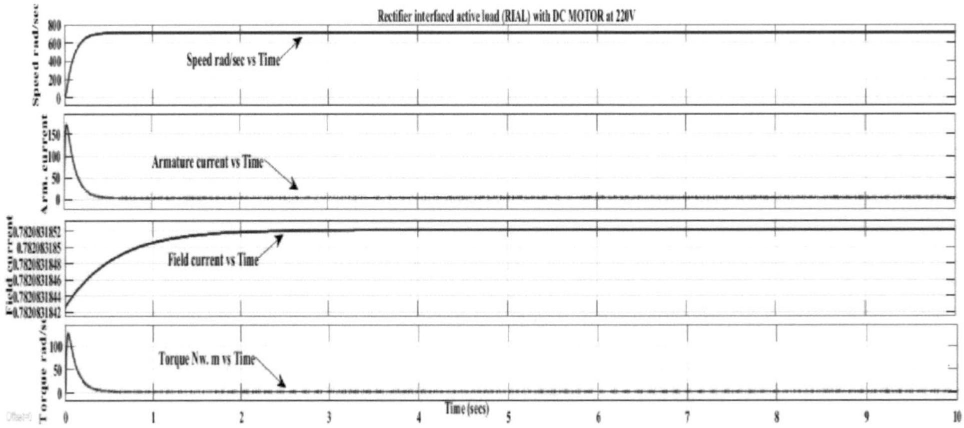

Figure 19.
DC motor characteristic curves with field voltage 220 volts.

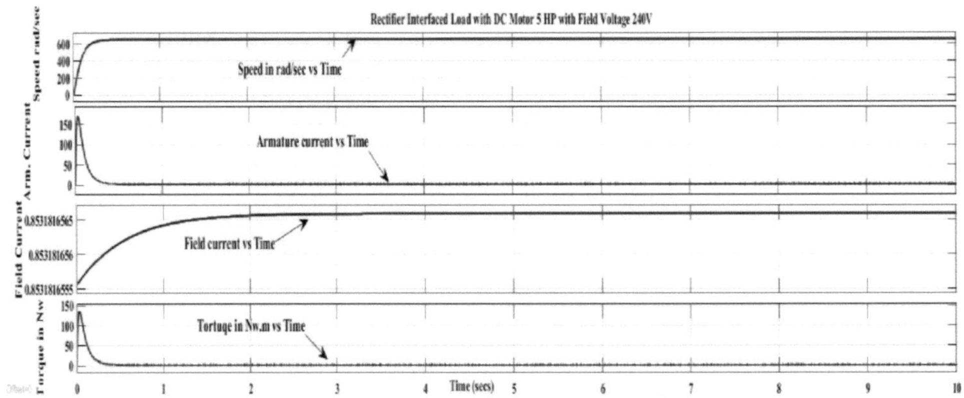

Figure 20.
DC motor characteristic curves with field voltage 240 volts.

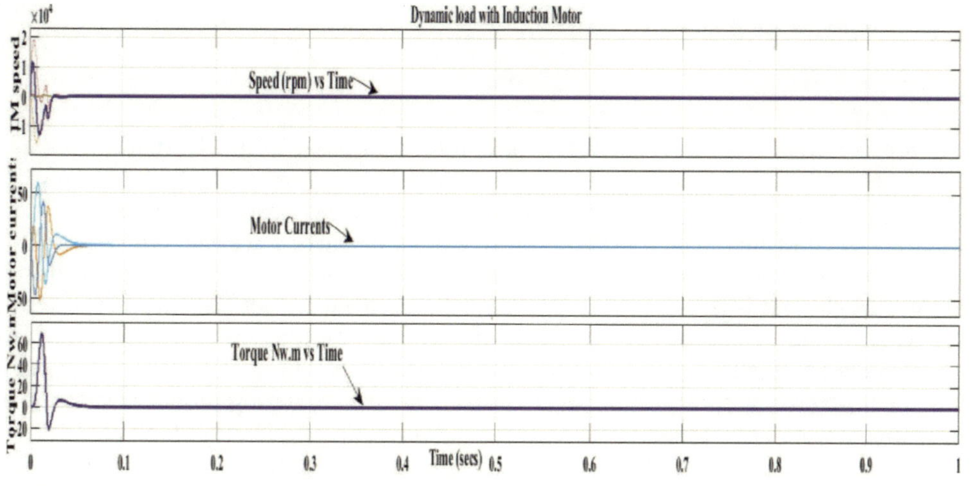

Figure 21.
Induction motor load with 10HP, 1500 rpm capacity.

by the motor currents characteristic curves in **Figure 21**. The controller frequency at PCC is well within the IEEE standard deviation between 49.6 and 50.1 Hz as shown in **Figure 22**. The voltage variations are also well within specified limits as per standards of IEEE, which are 365 and 410 volts (for nominal volts 415) as shown in **Figure 23**.

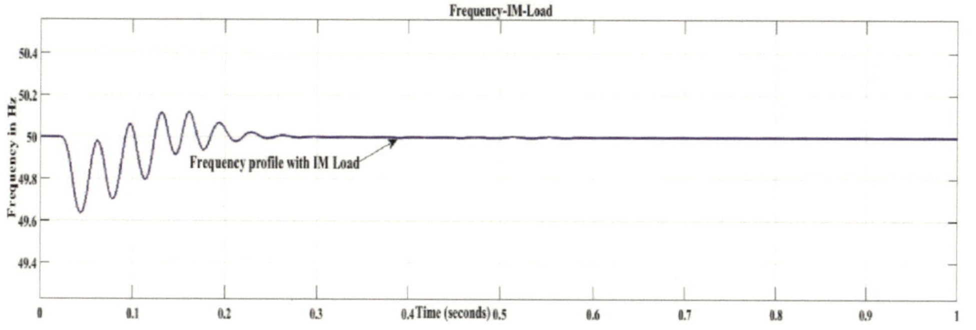

Figure 22.
Frequency profile with induction motor load 10HP, 1500 rpm capacity.

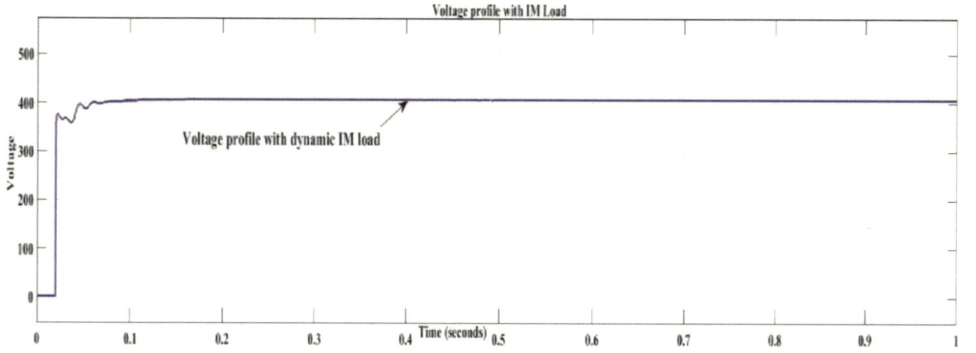

Figure 23.
Frequency profile with induction motor load 10HP, 1500 rpm capacity.

6. Conclusions and future work

6.1 Conclusions

Adoptive inverter control for stable operation of Microgrid in islanded mode has been presented in this chapter. A state space model for inverter interfaced Microgrid in a DQ reference frame with static and dynamic loads has been derived and used in designing of controller. The modeling is done in three parts. Part one, consists of DG module, part two network module and part three load model. The load model considered in this chapter is passive load, active load consisting of rectifier interfaced active load (RIAL), and dynamic load consisting of induction motor load (IM). The complete DQ model is obtained by mapping all sub-modules of DG, network, and loads, which consists of passive, active, and dynamic loads. The total mapping model developed in DQ frame is tested for dynamics of frequency, network, and load with all control loops. The time domain simulation results validated the linearized model. The MATLAB/Simulink simulation results prove the efficiency and effectiveness of the adoptive controller designed, in respect of Microgrid in islanded mode.

6.2 Future work

The work can be extended with hybrid renewables such as wind turbines, fuel cells, and energy storage systems (ESS), to use as DGs. The proportional load sharing with DG interfaced inverter controllers can be studied. Also, the stability of the total system can be assessed and controller's control can be modified to obtain absolute stability by plotting poles and zeros of characteristic equation in real and imaginary axes. With this, the total stability of the Microgrid can be obtained in islanded mode.

Acknowledgements

The authors acknowledge with thanks the management and faculty of Electrical and Electronics Engineering Department of Koneru Lakshmaiah Education Institute, Guntur-522302, A.P., India, Vasavi Engineering College, Hyderabad-500031, India and Datta Meghe Institute of Engineering Technology and Research, Wardha 442004, India.

Author contributions

This research was a collaborative effort between all the participating authors with equal contributions.

Conflict of interest

The authors declare that there is no conflict of interest.

Financial disclosure

The authors declare that this work has not received any financial aid from any source.

Appendices

v_i	inverter output voltage
v_i	inverter reference input voltage
Δ	incremental value
$\mathbf{K_p}$	controller proportional constant
$\mathbf{K_i}$	controller integral constant
ω_n	system normal frequency
m-bus	microgrid bus
g-bus	grid bus
T_e	is motor torque
T_L	is load torque
r_S	induction motor stator resistance
r_n	virtual resistor of large value, between each node and ground
$\mathbf{L_m}$	induction motor stator resistance
ω	induction motor stator angular frequency
s	induction motor rotor slip
$\mathbf{L_{ss}}$	induction motor stator inductance
α	delay angle of controller rectifier

Nomenclature

DER	distributed energy resources
DG	distributed generation
RES	renewable energy sources
MG	microgrid
PCC	point of common coupling
abc	time variable sinusoidal a, b, c quantities
DQ	network direct and quadrature axes quantities
dq	load side direct and quadrature axes quantities
RL load	resistive and inductive load
CPL load	constant Power Load
RIAL load	rectifier interfaced active load
IM load	induction motor load
LPF	low pass filter
LCL filter	inductance, capacitance and line side inductance
PI	proportional integral
AC	alternating current
DC	direct current
P	active power
Q	reactive power
UPS	uninterrupted power supply
PV	photo voltaic
VSI	voltage source inverter
AL	active load
PL	passive load
GHG	green house gases
abc to dq	park transformation, transforming abc quantities to dc rotating

Author details

Bangar Raju Lingampalli[1]*, Subba Rao Kotamraju[1], Kiran Kumar Malligunta[1], Bishnu Prasad Muni[2] and Rahul S. Somalwar[3]

1 Koneru Lakshmaiah Education Foundation, Guntur, A.P., India

2 Vasavi Engineering College, Hyderabad, India

3 Bajaj Institute of Technology, Wardha, India

*Address all correspondence to: lsmlbr@yahoo.in

References

[1] Vasquez JC, Guerrero JM, Luna A, Rodríguez P, Teodorescu R. Adaptive droop control applied to voltage-source inverters operating in grid-connected and islanded modes. IEEE Transactions on Industrial Electronics. 2009;**56**(10): 4088-4096. DOI: 10.1109/TIE.2009.2027921

[2] Zhang L, Wang X, Zhang Z, Cui Y, Lin L, Cai G. An adaptative control strategy for interfacing converter of hybrid microgrid based on improved virtual synchronous generator. IET Renewable Power Generation. 2021, 2022;**16**:261-273. DOI: 10.1049/rpg2.12293

[3] Chen Z, Luo A, Wang H, Chen Y, Li M, Huang Y. Adaptive sliding-mode voltage control for inverter operating in islanded mode in microgrid. Electrical Power and Energy Systems. 2015; **66**(2015):133-143. DOI: 10.1016/j.ijepes.2014.10.054

[4] Salem Q, Alzaareer K. Detailed analysis of grid connected and islanded operation modes based on P/U and Q/f droop characteristics. International Journal of Power Electronics and Drive System (IJPEDS). 2021;**12**(2):772-782. DOI: 10.11591/ijpeds.v12.i2

[5] Dashtdar M, Flah A, Sadegh SM, Hosseinimoghadam, El-Fergany A. Frequency control of the islanded microgrid including energy storage using soft computing. Scientific Reports. 2022;**12**: 20409. DOI: 10.1038/s41598-022-24758-6

[6] Fani B, Shahgholian G, Alhelou HH, Siano P. Inverter-based islanded microgrid: A review on technologies and control. Advances in Electrical Engineering, Electronics and Energy. 2022;**2**:100068. DOI: 10.1016/j.prime.2022.100068

[7] Hammer B, Gong K, Konigorski U. Modeling and control of inverter-based microgrids. IFAC Papers OnLine. 2018; **51**(2):19-24. DOI: 10.1016/j.ifacol.2018.03.004

[8] Hosseinimoghadam SMS, Roghanian H, Dashtdar M, Razavi SM. Power-sharing control in an Islanded microgrid using virtual impedance. In: 8th IEEE International Conference on Smart Grid. Paris, France: Research Gate IEEE Publications; 17-19 June 2020. pp. 1-5. DOI: 10.1109/icSmartGrid49881.2020.9144802

[9] Tucci M, Ferrari-Trecate G. Voltage and frequency control in AC islanded microgrids: A scalable, line-independent design algorithm. IFAC Papers OnLine. 2017;**50**(1):13922-13927. DOI: 10.1016/j.ifacol.2017.08.2212

[10] Meegahapola L, Laverty D, Jacobsen M-R. Synchronous islanded operation of an inverter interfaced renewable rich microgrid using synchrophasors. IET Renewable Power Generation. 2017;**12**:407-414. DOI: 10.1049/iet-rpg.2017.0406

[11] Ghazzali M, Haloua M, Giri F. Modeling and adaptive control and power sharing in islanded AC microgrids. International Journal of Control, Automation and Systems. 2020; **18**:1229-1241. DOI: 10.1007/s12555-019-0222-2

[12] Pham XHT. An improved controller for reactive power sharing in islanded microgrid. Electrical Engineering. 2021; **103**:1679-1689. DOI: 10.1007/s00202-020-01160-x

[13] Roudbari ES, Beheshti MTH, Rakhtala SM. Voltage and frequency

regulation in an islanded microgrid with PEM fuel cell based on a fuzzy logic voltage control and adaptive droop control. IET Power Electronics. 2020; **13**(1):78-85. DOI: 10.1049/iet-pel.2019.0592

[14] Naderipour A, Abdul-Malek Z, Davoodkhani IF. Load-frequency control in an islanded microgrid PV/WT/FC/ESS using an optimal self-tuning fractional-order fuzzy controller. Environmental Science and Pollution Research. 2023;**30**: 71677-71688. DOI: 10.1007/s11356-021-14799-1

[15] Paylakhi SZ, Pariz N, Bayaz MHJD. Distributed secondary control of islanded microgrids using modified integral backstepping technique. International Journal of Dynamics and Control. 2022;**10**:2011-2019. DOI: 10.1007/s40435-022-00917-5

[16] Su H, Zhang Z, Wang S. Island microgrid power control system based on adaptive virtual impedance. Frontiers in Energy Research. 2023;**10**:974288. DOI: 10.3389/fenrg.2022.974288

[17] Mi Y, Yan L, Cai H, Fu Y, Tian S, Jin C. Reactive power sharing control for islanded AC microgrid based on fuzzy adaptive compensation method. IET Generation Transmission and Distribution. 2020;**14**:585-593. DOI: 10.1049/iet-gtd.2019.0668

[18] He J, Li YW. Analysis, design, and implementation of virtual impedance for power electronics interfaced distributed generation. IEEE Transactions on Industry Applications. 2011;**47**(6): 2525-2538. DOI: 10.1109/TIA.2011.2168592

[19] Zandi F, Fani B, Sadeghkhani I, Orakzadeh A. Adaptive complex virtual impedance control scheme for accurate reactive power sharing of inverter interfaced autonomous microgrids. IET Generation Transmission and Distribution. 2018;**12**:6021-6032. DOI: 10.1049/iet-gtd.2018.5123

[20] Zhao C, Sun W, Wang J. Distributed robust secondary voltage control for islanded microgrid with nonuniform time delays. Electrical Engineering. 2021; **103**:2625-2635. DOI: 10.1007/s00202-021-01251-3

[21] Karimi H, Beheshti MTH, Ramezani A, Zareipour H. Intelligent control of islanded AC microgrids based on adaptive neuro-fuzzy inference system. Electrical Power and Energy Systems. 2021;**133**(2021):107161. DOI: 10.1016/j.ijepes.2021.107161

[22] Sen S, Kumar V. Microgrid control: A comprehensive survey. Annual Reviews in Control. 2018;**45**:118-151. DOI: 10.1016/j.arcontrol.2018.04.012

[23] Chen M-R, Wang H, Zeng G-Q, Dai Y-X, Bi D-Q. Optimal P-Q control of grid-connected inverters in a microgrid based on adaptive population extremal optimization. Energies. 2018;**11**(8):2107. DOI: 10.3390/en11082107

[24] Shah AA, Han X, Armghan H, Almani AA. A nonlinear integral backstepping controller to regulate the voltage and frequency of an islanded microgrid inverter. Electronics. 2021;**10**: 660. DOI: 10.3390/electronics10060660

[25] Sun J, Hai G, Ziabari MT, Mousavi H. Secondary control of islanded microgrid using robust finite time decentralized method based on adaptive fuzzy sliding mode controller. Systems Science & Control Engineering. 2021;**9**(1):714-723. DOI: 10.1080/21642583.2021.1989634

[26] Davari M, Jiang WGZP, Lewis FL. An optimal primary frequency control

based on adaptive dynamic programming for islanded modernized microgrids. IEEE Transactions on Automation Science and Engineering. 2021;**18**(3):1109-1121. DOI: 10.1109/TASE.2020.2996160

[27] Elnady A, Nasir AM. Efficient adaptive controllers and recursive filters for centralized control scheme of islanded microgrid with linear and non-linear loads. International Journal of Electrical Power & Energy Systems. 2022;**141**:108104. ISSN 0142-0615. DOI: 10.1016/j.ijepes.2022.108104

[28] Ndiwulu GW, De Jaeger E, Lusala AK. Islanded microgrid voltage control structure small signal stability analysis. In: IEEE Conference Paper, Milan, Italy. 2020. pp. 1-6. DOI: 10.1109/PTC.2019.8810896

[29] Kim J, Guerrero JM, Rodriguez P, Teodorescu R, Nam K. Mode adaptive droop control with virtual output impedances for an inverter-based flexible AC microgrid. IEEE Transactions on Power Electronics. 2011; **26**(3):689-701. DOI: 10.1109/TPEL.2010.2091685

[30] Jha SK, Kumar D, Tripathi PR, Appasani B, Zawbaa HM, Kamel S. An adaptive vector control method for inverter-based stand-alone microgrids considering voltage reduction and load shedding schemes. IET Generation Transmission and Distribution. 2023;**17**: 2703-2717. DOI: 10.1049/gtd2.12839

[31] Saifudheen P, Thresia MM. A droop controller based active power sharing of parallel inverter islanded microgrid. In: Conference: 2022 International Conference on Futuristic Technologies in Control Systems & Renewable Energy (ICFCR). Springer Nature Singapore Pte Ltd; 2022. pp. 1-10. DOI: 10.1109/ICFCR54831.2022.9893658

[32] Lou G, Wei G, Wang L, Xu B, Wu M, Sheng W. Decentralised secondary voltage and frequency control scheme for islanded microgrid based on adaptive state estimator. IET Generation Transmission and Distribution. 2017; **11**(15):3683-3693. DOI: 10.1049/iet-gtd.2016.1910

[33] Yang C, Ni S, Dai Y, Huang X, Zhang D. Anti-disturbance finite-time adaptive sliding mode backstepping control for PV inverter in master–slave-organized islanded microgrid. Energies. 2022;**2020**(13):4490. DOI: 10.3390/en13174490

[34] Le Minh P, Due DHV, Xuan HPT, Minh HN. A new control strategy of power sharing in islanded microgrids. In: 2017 International Conference on System Science and Engineering (ICSSE). 2017. pp. 175-179. DOI: 10.1109/ICSSE.2017.8030860

[35] Ghodsi MR, Tavakoli A, Samanfar A. A robust controller design for load frequency control in islanded microgrid clusters. Hindawi International Transactions on Electrical Energy Systems. 2022;**2022**:4218067. 12 pages. DOI: 10.1155/2022/4218067

Chapter 7

Communication Concept in Smart Grid Using Internet of Things

Anna Jarosz

Abstract

The use of distributed energy sources in combination with energy storage is gaining widespread attention due to ensuring the continuity of power supply. Artificial intelligence is used to control such a system effectively. This manuscript sheds light on how the communication challenges of the smart grid. Here, the radical and exciting conception based on Internet of Things (IoT) for sharing data information is presented.

Keywords: communication, smart grid, internet of things, artificial intelligence, microgrid

1. Introduction

This research on smart grid communication systems based on wireless techniques has a short tradition. As this field is maturing, amounts of methods using algorithms are becoming more and more understandable. A challenging problem in this domain is delivering the required energy supplies to connected loads simultaneously using the available power from renewable energy sources. This problem has attracted more attention in the field of artificial intelligence.

One of the most innovative ways of tackling the issue of communication in smart grid systems is the application of Internet of Things modules. One approach to solve this problem involves using a wide range of frequencies to send data to platforms. This technique enables communication and results from the basic functionality of automatic control. To illuminate the uncharted areas in smart grid systems, systems in this field are indicated with a detailed discussion of the processes occurring in the operation of microgrids. In addition, the functionalities of artificial intelligence in metering data collection were defined.

The critical elements in the tested structure necessary for the system's proper operation were also indicated. The result of the research is the methodology of controlling the communication system using the Internet of Things. In addition, an invented method aimed at sharing and analyzing the flowing data allows for an efficient energy supply to the receivers.

The chapter begins with an introduction, providing an overview of the topic and highlighting the importance of networking and communications solutions in the context of the Internet of Things (IoT) within the smart grid. Next, the chapter conducts a comprehensive literature review, exploring the existing body of knowledge on networking and communications solutions for IoT in the smart grid. It showcases the practical implementation of networking and communications solutions in a real-world scenario, offering valuable insights and examples from this case study. It delves into the role of AI in optimizing grid operations, improving efficiency, and enabling intelligent decision-making. This section examines the role of Information and Communication Technology (ICT) systems in supporting networking and communications solutions for IoT in the smart grid. It addresses the challenges and opportunities associated with these communication systems, emphasizing their impact on grid connectivity and efficiency.

This section provides a comprehensive overview of the IoT system concept within the smart grid context. It examines factors such as security, interoperability, scalability, and reliability, offering insights into potential mitigation strategies. The discussion section serves as a platform for the authors to analyze and interpret the findings. The chapter concludes by summarizing the key points discussed and highlighting the main takeaways.

2. Literature review

An ever-increasing body of literature shows that smart grids are being studied more and more. Also, in recent years, there has been considerable interest in implementing artificial intelligence into standard systems, e.g., medical decision support systems [1].

Ref. [2] reviewed a relatively new area of research connected with the application of Internet of Things modules. The idea is that monitoring and notifying the occurrence of a fault can be done by the developed system that operates on solar and wind energy sources.

Initial studies on hybrid microgrids supported by artificial intelligence primarily focused on the import and export of solar power, as shown in [3]. Syncing information in the proposed database system is up to be a provider of reports with simultaneous notification. This paper aims to create a visualization data system that makes analysis.

Ref. [4] Projected estimation method in minimization of power losses in distribution and transmission infrastructures. The invented idea uses the assumptions of a smart grid, which combines the possibilities of using Internet of Things modules. Another aspect of microgrids to consider is a storage system due to the insurance of renewable energy sources.

Studies by Minai et al. [5] have led to a more profound understanding of Internet of Things possibilities. The battery charges consider using artificial intelligence to transfer the surplus energy. It can be used to power demand side power management.

The problem of maintaining security in smart grid systems was discussed by Barman et al. [6]. The author employed an AI methodology that prescribes IoT modules' usage to detect power theft. In addition, the proposed system enables producers and consumers to view uploaded information taken from artificial intelligence-based devices.

Different approaches to managing protection systems aim to identify transmission line faults in the field of smart grids Pradyumna, [7]. The whole process can be done using an IoT-based two-way connection between producer and consumer by periodically monitoring and checking.

Parameters Farid et al. [8] studies is an area of research that emerged from the need for data analysis. The consistent use of IoT can also create a platform for collecting exported data in energy management systems. The architecture of IoT systems was presented, and the services offered to users were distinguished.

3. Smart grid case

A Smart Grid is considered to be an electricity-delivery network based on digital technology. It can supply electrical power to loads with the usage of two-way communication. The system can monitor the energy assessment unit's status and power delivery through the smart grid's delivery networks [9]. It can assist in calculating the electrical energy demand by monitoring and identifying degradation to the appliances, which can also help reduce maintenance costs. Furthermore, it can transmit and receive information through the smart grid so that people can know the status of their instruments in different time zones. This helps the healthy management of the power supply chain so that the system works properly.

The smart grid will enable citizens to participate actively in energy provisioning. Additionally, it will accommodate all generation and storage options, as well as facilitate new energy product and service markets, optimize existing asset utilization, offer efficient power quality for digital businesses and smart devices, keep pace with changing prices, and operate resiliently to attack and natural disaster,

Frequency stability and control are among the most critical problems in smart grid design and performance. Control loops try to keep the electricity grid parameters close to their reference value [10]. Each has its regulation component, ensuring stability (low-frequency or high-frequency control). With proper operation, there are critical issues in power system projects and processes. Most of the balance on the demand-supply side is achieved by controlling the output of distributed generating units.

Another crucial parameter to control is voltage. This is one of the leading smart grid operational challenges due to the high use of renewable and distributed sources. Alternative energy sources depending on weather conditions, such as wind speed and insolation, significantly change the voltage profile in the network [11]. As a result, combined generation and load profiles can lead to high system losses due to undervoltage and overvoltage.

Considering [the Smart Grid Reference Architecture working group report], the infrastructure of the studied system comprises five recurrent levels that reflect various goals and procedures as well as roles, data modeling, communication protocols, and elements. The compatibility categories were introduced in the Reference Architecture workgroup review and represented by five levels, as shown in **Figure 1**. Each layer contains at least two elements referring to smart grid operations.

The representation proposed by CENELEC [8] includes domains in the context of the smart grid architecture, showing the generation, transmission, and distribution of energy from power sources to loads. At the same time, zones represent the hierarchical structure, including components of each domain, as **Figure 2** presents.

Generating refers to producing vast quantities of electric power from fossil, hydro, and nuclear power plants, local photovoltaic power systems, and offshore wind farms. Transmission is understood as a representation of the structure and group responsible for long-distance power transportation. Finally, the distributing estate process and entity responsible for supplying power to consumers.

Power sources may be distributed electrical resources are small-scale power generation systems directly linked to the standard electric grid (generally in the range of 3 to 10.000 kW).

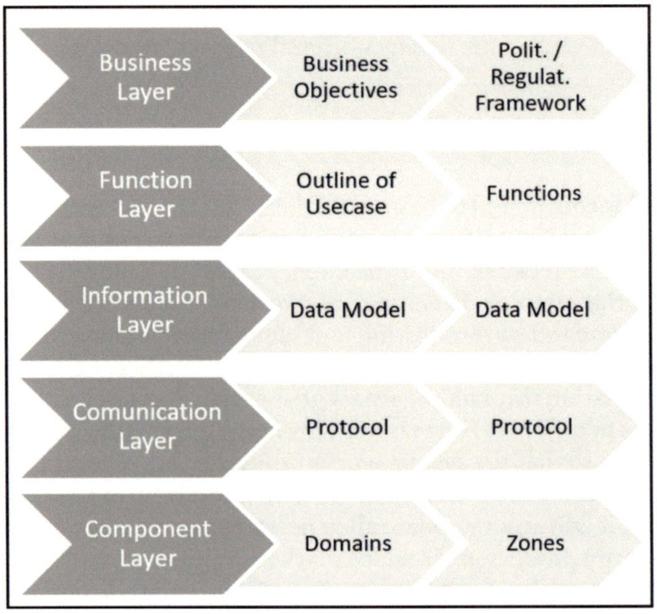

Figure 1.
Smart grid layers due to the Canelec concept [12].

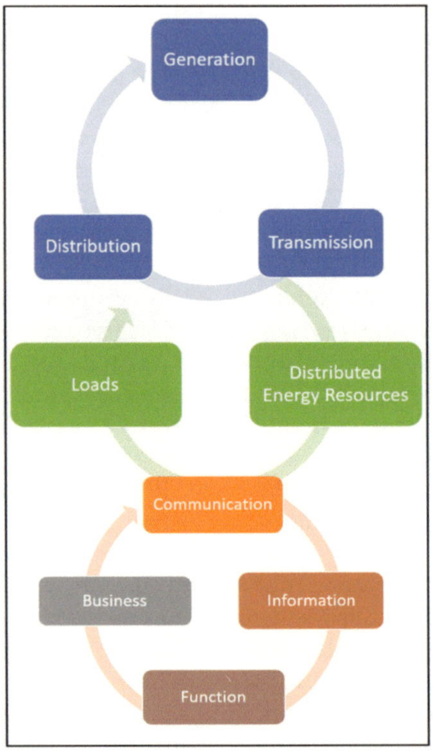

Figure 2.
Smart grid structure with connections [12].

Both end users and producers of electricity are housed on customer premises. The property includes a commercial, residential, and industrial structure. Additionally, energy is hosted, including power panels, batteries, and stores for electric vehicles and microturbines.

4. Artificial intelligence in smart system

The Smart Grid Architecture is an electrical grid model that aims to understand how electricity is connected to the energy network. In addition, the model seeks to portray expansion, the installation of the electric grid, and the transition to new technologies and methods for the smart grid.

Artificial intelligence is used to control such a system effectively. This article sheds light on the communication challenges of the smart grid [13]. Here, the radical and exciting conception based on IoT for sharing data information is presented.

The Smart Grid Architecture is a model of the electrical grid that aims to provide an understanding of how electricity is connected to the energy network. In addition, the model aims to portray the installation of the electric grid and the transition to new technologies and methods for the smart grid.

Database systems 'migrate' across remarkably various information structures at multiple places and with diverse ownerships because services delivered by smart grid (i.e., general application case) stretch over many areas/regions. As a result, it is advised that the data sets should be classified and tagged.

The diversity and richness of the Smart Grid have reflected as well as the many cell (area/region) specificities, which highlight how challenging it is to handle the integrity of the Smart Grid in its entirety. All relevant organizational and technological elements must be considered for each cell. Additionally, when a use case spans multiple cells, issues are made even more complicated.

According to the Ministry of Power in India, the term Smart Grid covers an appropriately automated power grid in which communication employing IT systems plays an important role—controlling the power flow by monitoring it in the area of the electricity generation zone and the location of its final consumption, taking into account the devices powered by energy. In smart grid systems, load limiting is also used to adjust the generated energy constantly in real time.

Smart grids can be achieved by implementing efficient transmission and distribution systems, system operations, consumer integration, and integration with renewable sources [4]. In addition, smart grid solutions help to monitor, measure, and control power flows in real-time, which can help to identify losses and thus take appropriate technical and management actions to stop losses.

With the use of smart communication systems, it is possible to obtain economic benefits resulting from the reduction of losses during peak load management and in systems performing the transmission and distribution of electricity, as depicted in the **Figure 3**.

In addition, sharing information on energy generation, demand, and supply can enable improved service quality, more appropriate asset management, and increased reliability of renewable energy integration. In the longer term, the use of artificial intelligence in power systems may create opportunities for the creation of self-healing networks.

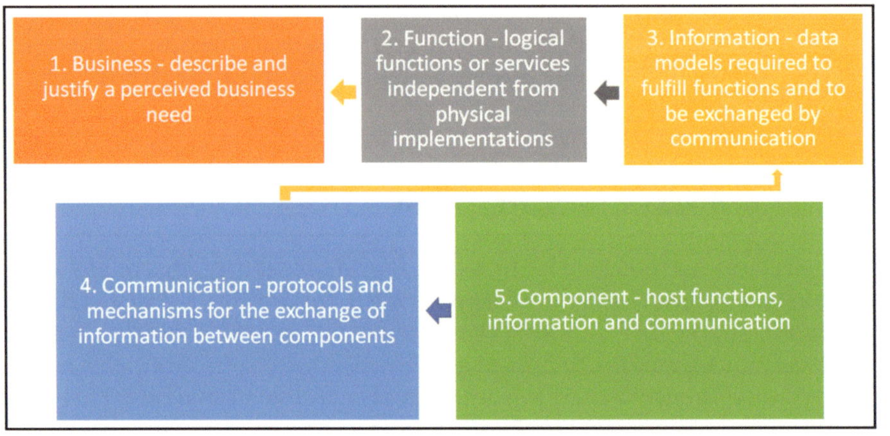

Figure 3.
Smart grid levels with primary operations [12].

The 'Internet of Things' refers to the network of connected anything. The term is intended to relate to Internet-based protocols, local vocabularies, standards, and application logic [14]. The changed networks then adopt the outlook of the Internet of Everything items throughout to conduct information sensing, exchanging, and communicating to realize smart registrations, locations, maps, and administrations.

Monitoring equipment performance and avoiding unnecessary maintenance using AI replaced the work of many people. Instead of employing routine human inspections, connected sensor technologies can offer a more thorough understanding of the technology that supports and is used in the smart grid [15]. Furthermore, to guarantee there is no waste, the equipment of IoT can be monitored, among other things, for voltage, frequency, and power.

5. ICT systems

Current communication channels must be used, and technological advances should be incorporated to some level depending on the importance and intensity of the operations. They may be used to decrease the expense of establishing such a smart information exchange.

Over the ensuing years, technological advances can increase power systems' connectivity, intelligence, efficiency, reliability, and sustainability as the technical transition between analog and digital progress.

Technologies for information and communication (ICT), current detectors, big data in machine learning, and IoT are among the techniques to help improve efficiency and reduce carbon emissions of energy infrastructure [16].

The ICT infrastructure is instrumental so that consumers can benefit from new ways to engage in the energy transition and better services based on digital innovations, more efficient energy use, and energy savings.

It is crucial to ensure that the ICT sector operates effectively and sustainably in terms of energy use and consumption. A future-proof architecture that includes shared protocols and reliable servers for future and current generations is required to create digital capabilities.

Appropriate protection, stability, security and moral principles must be maintained in the face of problems brought on by digitalization and the widespread processing of data, especially concerning cyber security concerns.

Various principles for energy efficiency are being developed as a consequence of the growing interaction between individuals and their electronics, what is known as the Internet of Things (IoT). In addition, it can positively affect the power industry, such as enhancing the oversight of energy flows in operating electric smart grids.

A revolutionary role for IoT [17] in power generation, improved productivity, accessibility, and dependability, can be seen in enhancing transparency in the transfer of energy subsidies. Furthermore, IoT may significantly contribute to energy conservation and savings by detecting and monitoring.

This AI-based technique could also enhance bandwidth utilization through systems that rely upon productivity growth brought on by more industrial automation, networks, and equipment. In addition, by allocating loads based on availability, data systems can promote behavioral modifications and lower emissions.

It may expand e-government projects, offering a communications infrastructure to boost the quality of the electric grid and structure reliability, mechanization, and extraordinary interconnection in transit systems, and collecting evidence of energy markers to evaluate advancement with recognition of highlighted issues.

Giving independent users more influence over systems of distributed generation like alternative energy sources, trying to raise customer perception of sustainability and green energy demand, and speeding up the creation of brand-new electricity systems that AI modules may operate.

To handle the fluctuating and decentralized character of renewable sources, IoT-enabling technologies are crucial for expanding the distributed energy sector. It can assist with improving utility performance and expense and integrating an approaching generation of sustainable energy sources. Moreover, further usage can be economically advantageous when coupled with demand-side control initiatives.

Without IoT-enabled proper monitoring, regulation, and supervision, it could be hard to penetrate the market for renewable energy sources further because they are irregular (air velocity for wind, irradiation for sun, and water flow are all unpredictable). Furthermore, to reduce the costs of evaluation, management, and maintenance, IoT enables the application of extensive data analysis [18] and convergence of consumption, delivery, and energy generation.

6. Internet of things-based communication

Services that are hard to accomplish with the use of wires, such as long-range communications, are made possible by wireless operations. The phrase is frequently utilized in telecommunication services to describe major platforms (such as transmitters and receivers, remote controllers, Internet technology, and connectivity) [19] that transmit information wirelessly using one or more types of energy, such as radio frequency (RF), infrared light, laser beam, light waves, sound energy, etc. This way, information is transmitted over billions of kilometers for radio communications and short distances (only a few meters using a television remote control).

Several IoT use cases include applications that require ultra-reliability, real-time connectivity, extremely low latency, extremely high availability, and assured in-time supply. For such applications, the unit-approving concept for spectrum access might be appropriate to control the level of service, avoid undesirable delays, and reduce interference.

Smart grids can gather information from their surroundings and operations and respond in real-time thanks to sensors. As they provide them with the signals needed to collect essential data, sensors are undoubtedly the foundation of smart systems. Moreover, to meet various application goals, several sensors can be implemented.

By identifying high-energy devices and taking immediate action to limit their consumption, smart devices, and electric connectors also enable more accurate energy consumption monitoring across a facility. In addition, IoT sensors can provide adequate energy management solutions when used to their maximum capacity, which is a challenge for system managers as energy prices rise and sustainability goals climb.

The significance of coordinating spectrum utilization across all services is undeniable, as is the growing challenge in achieving this shared objective as spectrum demand rises. For successful communications, spectrum allocation and usage must be coordinated. According to the author in [5], the multi-stakeholder character of spectrum use relates to requests on a finite resource necessary for services to assure, among other things, socioeconomic development, worldwide coverage, and life safety. Moreover, innovative services and solutions require an adequate spectrum in suitable frequency ranges.

It is essential to consider that technology, depending on the scope, will be beneficial in achieving the UN Sustainable Development Goals, reducing the consequences of climate change, and moving toward sustainable human development worldwide. As unpredictable weather phenomena occur widely, monitoring and controlling the environment, weather forecasting, and related services are more crucial than ever. These services also assist in the security of individuals and properties, improving continuous improvement and reducing global warming.

The energy industry (utility services and network operators) highlighted the need for technological advances (LTE-450, LTE-M, and NB-IoT) to function in the 450 MHz frequency range. Also, it is indicated in [20] that the tendency for 'smart meter' application in the electrical sector is viewed as 'critical communication'. Therefore, it will be necessary to significantly enhance data rates and spectrum access to implement techniques based on radio spectrum to smart grids.

Relying on the GSM/EDGE and UMTS/HSPA specifications, LTE, which is Long-Term Evolution, is a communication protocol for portable devices and data centers. It enhances the capacity and performance over those specifications by leveraging a radio broadcasting configuration and network infrastructure upgrades.

The low-powered wireless technology known as NB, or NarrowBand, was created from scratch for large-area machine-to-machine (M2M) connectivity. The wireless connections standard was designed to enable the most stable, power-efficient, and trustworthy sensing interaction.

One prominent LPWAN technology is LoRa (Long Range), which utilizes a low-power, wide-area modulation technique to enable long-range communication with low power consumption. LoRaWAN is the corresponding network protocol that enables secure and efficient communication between LoRa devices and gateways. LoRaWAN has gained popularity for its ability to provide long-range coverage, deep penetration, and low power consumption, making it suitable for applications such as smart cities, agriculture, asset tracking, and environmental monitoring.

Radio Spectrum Policy Group from the Directorate-General for Communications Networks, Content and Technology EC unit [21] proposes a roadmap to access the IoT spectrum in Europe, reflecting different uses and scenarios. Enhanced spectrum harmonization and intervention must be undertaken with caution, especially as it is not cost-effective and needs to take into account the other national circumstances of each country involved. The roadmap recommends focusing on specific types of spectrum bands, which can lead to the more cost-effective use of those bands at the national level.

Today nearly every object on the planet is connected to the Internet to receive and send data. Another related information is that most such devices now have connectivity to advanced wireless communication protocols. For example, we are now able to launch data using protocols, including some data via radio frequency, near-field technologies, Bluetooth, Wi-Fi and radio frequency identification (RFID).

Historically, communication technologies began to develop due to the discovery of electromagnetic waves and the study of their potential. This enabled the use of radio frequencies since 1901, when Guglielmo Marconi transmitted the first intercontinental signal wirelessly, linking North America to Europe [22]. However, the rules of the radio signal and the impact of the atmosphere on radio communications were still a great deal to be discovered. As a result, Marconi played a significant role in radio studies and advancements over the following 30 years.

With the development of technology, a rocket was built that launched the first spacecraft in 1962 [23]. After that, it became a telecommunications satellite that enabled the transmission of communication signals actively to distant places on earth. Radio technology has continued to be developed into a form that allows tracking tags to be attached to objects and automatically identify them [24]. The Pacific NW Bell telecommunications company created a type of wireless phone in the same year that connected the device to a radio transmitter, which attached customers to operators.

In 1983, an object communicating using RFID (Radio Frequency Identification) technology was registered in the Patent Office. This technology is based on offering distinctive identification and backend integration that enables various applications. Multiple parts make up each RIFD electronic system: a scanning transmitter, a receiver, and a reader.

An RFID transponder or analyzer is used when the scanning transmitter and receiver are integrated. Fixed readers and portable readers are the two different categories of RFID transponders, which are network-connected gadgets that can be carried about or set to a surface. It sends information that turns on the tags using electromagnetic radiation. After being turned on, the label returns a signal to the transmitter, which is converted into information.

In the meantime, mobile phones began to be developed, and one of the technology providers, Ericsson, developed Bluetooth radio technology in 1994 [25]. The BluetoothSpecial Interest Group was then formed, which focused on creating an open specification for small-area wireless data transmission for short-range communications, including between computers.

WLAN (Wireless Local Area Network) was introduced to connect many devices simultaneously in 1997 [26]. Based on the IEEE 802.11 standard, the Wi-Fi protocol was defined, allowing computers to connect wirelessly in a locally limited area. However, it described technology that increased our data rates and gave us more excellent channels but did not address efficiency. As a result, the Wi-Fi traffic congestion prevents efficient use of the medium despite the more excellent data rates and 40/80/160 MHz channels used by 802.11n/ac devices.

The continuation of the development of RFID was a particular category named Near-Field Communication (NFC) that allows communication close, invented by NXP Semiconductors and Sony in 2002 and introduced in 2004 [27]. When used by devices close to one another, NFC allows contact or contactless information transmission within a range of one to four centimeters and up to 10 units. While technologies like Wi-Fi and Bluetooth concentrate on radio broadcasts, NFC uses electromagnetic radio fields.

5G is a cutting-edge technology that represents the latest advancement in mobile communication systems. It is designed to provide faster data transfer speeds, lower latency, and increased network capacity compared to its predecessors. While 5G

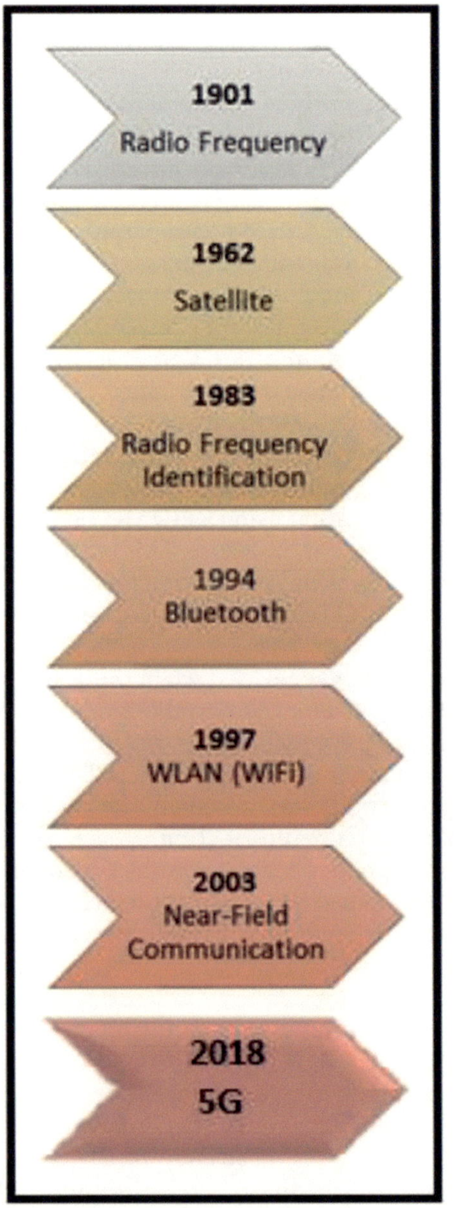

Figure 4.
Protocols supporting the internet of things modules with year implementation.

is commonly associated with improved mobile connectivity for smartphones and tablets, its potential extends far beyond that.

One of the significant advantages of 5G is its ability to support a wide range of applications, including IoT. With 5G, the IoT can reach new heights, enabling what is known as massive IoT [28]. This means that 5G networks can support a massive number of connected devices simultaneously, allowing for seamless communication and data exchange between them.

Figure 4 shows a timeline indicating the emergence of protocols supporting Internet of Things communication.

7. Internet of things system concept in smart grid

We advised starting with a particular issue, learning how the collected information may assist a Smart Grid system in making choices, and then tackling one other potential area when developing the IoT system for Smart Grid.

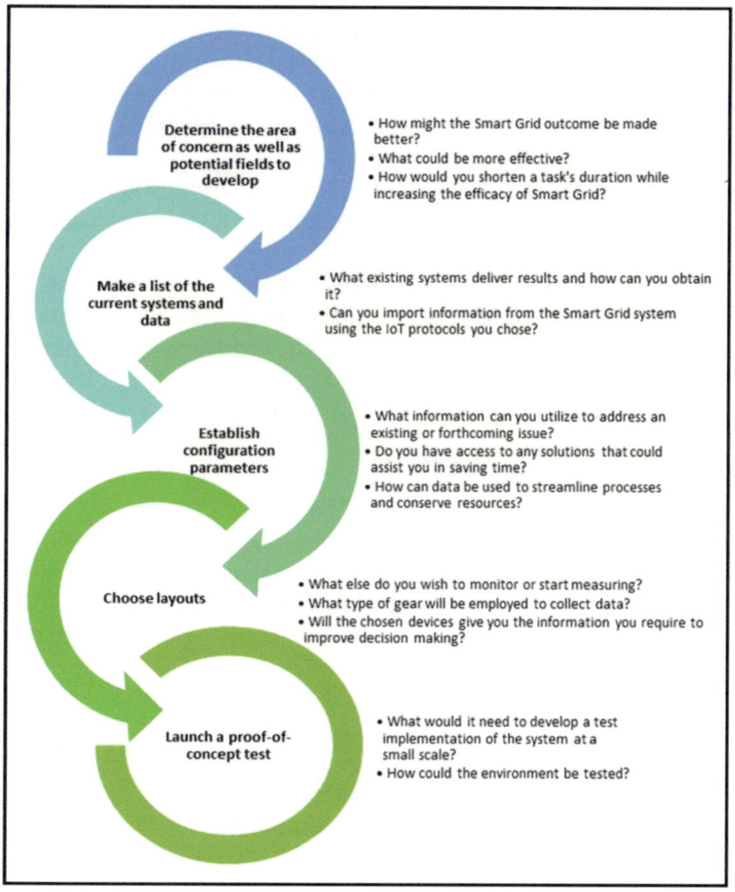

Figure 5.
Steps toward implementation of IoT-based communication system.

The first step in creating a smart system solution is to identify areas of improvement, catalog the current infrastructure, and develop the additional equipment needed to get started.

It is a beneficial idea to begin by posing some questions. To project an efficient IoT-based communication system for the smart grid, we chose several areas concerned. We selected several areas, including establishing measurements, recognizing the issue or opportunity, starting solid evidence, choosing patterns, and making an itemized list of the current systems and information.

There are a few established standard ways for setting up a Smart Gird. In almost any location, it can emerge. Despite the variations, several standard procedures have to be taken into account.

It is essential to consider maintenance while you develop a smart grid solution, set up new equipment, or introduce new communication technologies [29]. You might not be able to resolve problems if it was involved in the creation of apps for smart systems.

You might add many devices to your Smart Grid system to build a network protocol for your IoT devices, increasing the area covered and exposure. However, you could significantly lower the surface area and susceptibility if the connected technology does not connect with your core network directly but rather with a gateway that reduces the attackable surface.

When planning cost estimates, it is vital to remember that while a strategy to manage Smart Grid system space might be implemented in months, a solution to increase security might take much longer. To be sure the IoT-based communication system is safe, time should be anticipated to create specialized hardware.

Once a course of action has been chosen, gather the aid of existing staff and outside collaborators to develop a verification that will allow you to validate the solution. Again, trying out technology in your context, gathering data, and assessing performance may determine goals.

Including determining areas of concern and potential fields to develop, making a list of the current system and data, establishing configuration parameters, choosing layouts and launching a proof-of-concept test [30]. **Figure 5** shows the five fields listed above with crucial questions in each area.

8. IoT communication issues

From defining IoT goals and using cases that are appropriate for the smart grid, choosing the required IoT elements that are appropriate for the used model, building, and prototyping with optional integration and IoT applications with some other cutting-edge innovations and applying necessary security measurements, there are still some challenges during IoT implementation.

Tools used for IoT analysis assist in evaluating the information gathered by IoT sensors to enable more informed decision-making. Although there are unique obstacles to IoT, general data analysis problems equally apply to IoT implementation.

Adding more sensors to your system raises the likelihood of cyberattacks, although the IoT initiatives offer multiple economic prospects. More IoT-connected devices are susceptible to cybersecurity breaches, according to research [2]. As a result, the main obstacle to the adoption of IoT protection.

IoT platform consists of various instruments, scanners, and gadgets, and every supplier tries to set the mainstream technology. Therefore, interoperability of IoT modules with current systems is essential for effective deployment. Integrated cloud solutions, a dearth of defined M2M communications, and differences in technology and data processing between IoT systems are some inconsistencies.

Although some IoT systems operate on AC power, such structure contains equipment employed in harsh environments and relies solely on their battery for power. Businesses may keep records of whether an IoT device's battery has to be charged or updated. Generate a responsible IoT system by discovering gadgets that create or preserve energy while not being used. Replacement can be difficult, particularly if a device is positioned in a challenging-to-access location.

When IoT technologies are deployed, the dataset expands rapidly. As a result, owners require powerful information processing technology, large-sized storage, and increasing data transfer to collect IoT data and execute analyses.

Unstructured data collected by IoT devices is challenging to analyze. For example, if the surroundings or components around the detectors are unstable, the information gathered may even have abnormalities. Such data quality problems must be found to enhance decision-making.

Artificial intelligence capabilities, including IoT, require high-quality and readily accessible data. Therefore, an operator needs to utilize the appropriate data sets and have a reliable supply of pertinent information that is fresh, available, well-governed, and protected to offer the most effective and up-to-date AI algorithms.

Large-scale data accessibility is required for most artificial intelligence technology solutions to develop strategies. However, even though producing enormous amounts of data opens up more business prospects, doing so also raises storage and safety concerns.

Needful information must be safeguarded from stealing or tampering, while artificial intelligence largely depends on knowledge for forecasts and recommendations.

Because AI systems are sophisticated, installing and training them takes time before they become suitable. Therefore, one of the main benefits of employing AI techniques is the implementation time delay.

Maintaining confidentiality, safety, and privacy in the communication system is necessary. Another challenge concerning IoT technology is using resources and spectrum, where the evaluation of dynamic architecture and network functions is needed [31].

A platform for communications requires wireless energy and data transfer integration to improve power generation and sharing. Creating remote access strategies with coding and sequencing may also be a rigid demand, as well as managing resources and interference.

Challenges during IoT-based communication system implementation can be sorted into six categories. The areas of issues are shown in **Figure 5**.

9. Discussion

The process of their validation based on a worldwide ISO standard is a crucial one that is currently at the beginning stage but is predicted to contribute considerably to the ongoing advancement of intelligent energy structures in the area of communications, as **Figure 6** illustrates. The development of sustainable power architecture must be built on recognized, interconnected solutions that guarantee the level of services rendered, security with electricity, and information assurance [32]. By taking assessment metrics into account and documenting the communications patterns, this procedure can provide the essential basis for ongoing advancements.

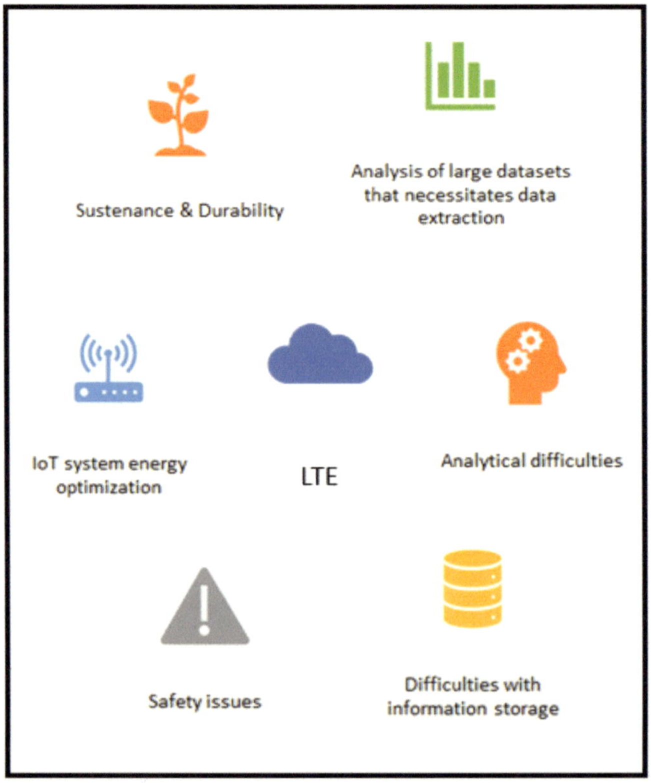

Figure 6.
Main fields of IoT-based communication system implementation issues.

For management services or specifications to be widely used and, correspondingly, for the entire agreement to be reached in the design, maintenance, and support of current energy regulations, it is essential to defeat competitive monopolistic efforts. To guarantee the continued deployment of smart grid technology, concerted efforts must be made to fix the problems and flaws discovered in the current norms and create brand-new ones [7]. This specification will offer a collection of rules, instructions, and methods to make sure that all essential components needed for an intelligent system to succeed are noticed.

The numerous options in the present rules and regulations that can define the layout, scheduling, sequence, management, and restoration of faults during information transmission make an overriding factor that poses an energy infrastructure challenge for designers [19]. This is to simplify the reality that existing standards of regulation cross over one another, causing major misunderstandings and operational inadequacies.

10. Protocol example for smart grid

First, it is important to identify the specific requirements of the smart grid system. This includes understanding the communication needs, data collection requirements, security considerations, interoperability requirements, and any specific constraints or limitations [33].

Next, key use cases and scenarios that the protocol needs to support should be defined. This involves identifying interactions between different components of the smart grid system, such as smart meters, sensors, controllers, and the control center for demand response.

The communication architecture for the smart grid system needs to be determined. This includes defining the layered architecture, components, and interfaces involved. Consideration should be given to the different communication technologies and protocols that may be used, such as LPWAN, SCADA protocols, or IoT protocols like MQTT or CoAP.

Designing the message formats is another crucial step. This involves specifying the data fields, encoding formats, and metadata for the messages exchanged between the smart grid components.

Based on the identified requirements, use cases, architecture, and message formats, the communication protocol for the smart grid system can be developed. Factors such as reliability, security, scalability, and interoperability should be considered during the design process. This may involve defining message exchange patterns, error-handling mechanisms, and synchronization protocols.

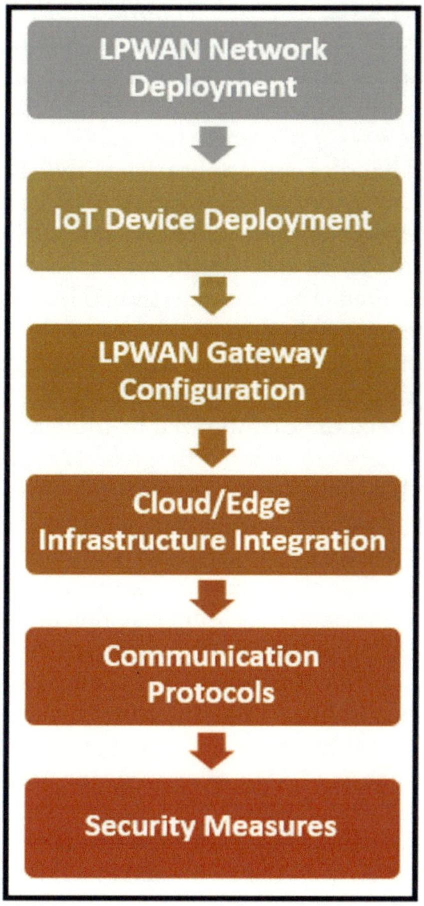

Figure 7.
Communication infrastructure with LPWAN deployment.

Once the protocol is designed, it needs to be implemented and tested. Software libraries should be developed to implement the protocol, and interactions between different components of the smart grid system should be tested through simulations. This ensures that the protocol meets the identified requirements and use cases.

Gathering feedback from stakeholders and experts is essential for refining the protocol. Their input can help identify areas for improvement and make necessary adjustments, updates, or enhancements to the protocol.

Formally documenting the protocol specifications and considering standardization is important for broader adoption and interoperability in the industry.

Lastly, it is crucial to continuously monitor the evolving needs and advancements in the smart grid domain. Staying updated with new technologies, security measures, and industry standards ensures that the protocol example remains relevant and effective over time.

By following these steps, a protocol example can be developed for a smart grid system that addresses the specific processes and requirements, enabling efficient and secure communication between different components and facilitating effective management and control of the grid.

To deploy a communication infrastructure with LPWAN in the smart grid, several steps need to be followed and they are shown in **Figure** 7.

First, the coverage area for the LPWAN network in the smart grid must be determined. This involves identifying areas that require connectivity and strategically deploying LPWAN gateways to provide comprehensive coverage.

Once the coverage area is established, the LPWAN gateways need to be installed and configured. These gateways will connect to the LPWAN network server, which manages the gateways and handles communication with IoT devices. Security protocols, encryption, and authentication mechanisms should be implemented to ensure secure communication and protect the network from unauthorized access.

Next, IoT devices specific to the smart grid, such as smart meters, sensors, and controllers, should be identified. These devices should be equipped with LPWAN communication modules compatible with the chosen LPWAN technology. They are then deployed across the smart grid infrastructure, considering factors like coverage, data collection requirements, and reliable connectivity.

The LPWAN gateways need to be configured to efficiently transmit data from the IoT devices to the LPWAN network server. This ensures a reliable flow of information throughout the network. Additionally, connectivity should be established between the LPWAN network server and cloud/edge computing resources. Data ingestion mechanisms are set up to transfer collected data to the cloud/edge infrastructure for further analysis and storage. The cloud/edge infrastructure should be configured to process, store, and analyze the data, providing valuable insights for decision-making.

To facilitate communication between IoT devices, gateways, and the LPWAN network server, the chosen LPWAN protocol (e.g., LoRaWAN or NB-IoT) should be implemented. This ensures efficient and secure communication. Compatibility and interoperability with other existing communication protocols and systems in the smart grid environment, such as SCADA or distribution management systems, should also be ensured.

Robust security measures, including encryption, authentication, access control, and intrusion detection systems, should be implemented to safeguard the communication infrastructure and protect transmitted data from potential threats. Regular updates and monitoring of security protocols are necessary to address emerging threats and vulnerabilities.

Integration mechanisms should be established to enable seamless data exchange and interoperability between the LPWAN communication infrastructure and existing systems and technologies in the smart grid. This ensures smooth coordination and data sharing between LPWAN-based IoT devices and other systems like SCADA or grid management platforms.

Finally, scalability and flexibility should be considered when designing the communication infrastructure. It should be able to accommodate the addition of new IoT devices and gateways as the smart grid evolves and grows. The infrastructure should also be flexible enough to adapt to future advancements in LPWAN technology and emerging requirements, allowing for the integration of new technologies seamlessly.

By following these steps, a robust and efficient communication infrastructure with LPWAN can be deployed in the smart grid, enabling effective data collection, analysis, and decision-making processes.

11. Conclusion

This article proposes innovative communication protocols for the unique problems of smart grid applications that must be considered while developing and implementing architectural designs. Due to the variability of structures and the changing nature of their area of operations, complexities must constantly be reduced, development works must be completed more quickly, and additional capacity must be added.

The supply of interconnected technical solutions that guarantee the accessibility of the electrical system parts and decrease the danger of different systems dropping in value is necessary for creating smart grids. Furthermore, to improve quality with strict policies, effective integration requires a straightforward, unambiguous approach to delivering end-to-end network connectivity rooted in proactive and compatible standards.

Author details

Anna Jarosz
Department of Power Electronics and Energy Control Systems, AGH University of Science and Technology, Cracow, Poland

*Address all correspondence to: anjarosz@agh.edu.pl

References

[1] Hamet P, Tremblay J. Artificial Intelligence in Medicine, PMID: 28126242. DOI: 10.1016/j.metabol.2017.01.011

[2] San P, Jha V, Srimat S, Das S, Abmash A. Design of smart grid and monitoring system using IoT. In: 2022 International Conference on Intelligent Controller and Computing for Smart Power (ICICCSP). IEEE Xplore; 2022. DOI: 10.1109/ICICCSP53532.2022.9862507

[3] Aagri D, Bishi A. Export and import of renewable energy by hybrid microgrid via IoT. In: 2018 3rd International Conference on Internet of Things: Smart Innovation and Usages (IoT-SIU). IEEE Xplore; 2018. DOI: 10.1109/IoT-SIU.2018.8519873

[4] Tauqir H, Habib A. Integration of IoT and smart grid to reduce Lin losses. In: 2019 2nd International Conference on Computing, Mathematics and Engineering Technologies (iCoMET). 2019. DOI: 10.1109/ICOMET.2019.8673433

[5] Minai A, Kumar S, Khan A, Kumar M. IoT based energy management system for smart grid. In: Conference ICACCM. IEEE Xplore; 2020. DOI: 10.1109/ICACCM50413.2020.9213061

[6] Barman B, Yadav S, Kumar S, Gope S. IoT based smart energy meter for efficient energy utilization in smart grid. In: 2018 2nd International Conference on Energy, Power and Environment: Towards Smart Technology (ICEPE). IEEE Xplore; 2018. DOI: 10.1109/EPETSG.2018.8658501

[7] Pradyumna A, Kuthadi S, Kumar A, Karuppiah N. IoT based smart grid communication with transmission line fault identification. In: International Conference on Intelligent Controller and Computing for Smart Power (ICICCSP). IEEE Xplore; 2022. DOI: 10.1109/ICICCSP53532.2022.9862383

[8] Farid N, Salmeh R. Network configurations for IoT Services in Smart Grid. In: 2021 9th International Conference on Smart Grid and Clean Energy Technologies (ICSGCE). 2021. DOI: 10.1109/ICSGCE52779.2021.9621636

[9] Elmrabet Z, Kaabouch N, Elghazi H. Cyber-Security in Smart Grid: Survey and Challenge's. Electrical Engineering Department. Computers & Electrical Engineering. 2018. p. 67. DOI: 10.1016/j.compeleceng.2018.01.015

[10] Irfan A, Hassan S, Ahmed R, Ishaq I. IoT based smart meter. In: 2019 International Conference on Innovative Computing (ICIC). IEEE; 2019. DOI: 10.1109/ICIC48496.2019.8966684

[11] Çelik D, Meral M. An overvoltage and undervoltage prevention control for superconducting magnetic energy storage based power converter connected to a weak grid. In: IEEE International Power and Renewable Energy Conference (IPRECON). IEEE; 2021. DOI: 10.1109/IPRECON52453.2021.9640990

[12] Smart Grid Information Security. CEN-CENELEC-ETSI Smart Grid Coordination Group; 2012

[13] Dilmegani C. IoT Implementation: Steps, Challenges, Best Practices in 2023. Journal of Business Research. AIMultiple; 2023;**161**. DOI: 10.1016/j.jbusres.2023.113838

[14] Patel K, Patel S. Internet of things-IoT: Definition, characteristics, architecture,

enabling technologies, application & future challenges. International Journal of Engineering Science and Computing. 2016;**6**:6122-6131

[15] Ihekoronye V, Nwakanma C, Oluchi A, Kim D. Benefits, challenges and practical concerns of IoT for smart manufacturing. In: 12th International Conference on ICT Convergence. DOI: 10.1109/ICTC52510.2021.9620771

[16] Modi V, Figueroa H. Sustainable Development Goal for Energy and Information and Communications Technologies. UN Chronicle; 2015;**52**(3):49-51. DOI: 10.18356/d7951adb-en

[17] Digitalising the energy system - EU action plan (COM(2022)552/2), Staff Working Document (SWD(2022)341)

[18] Saxena D, Rathor S. Energy management system for smart grid: An overview and key issues. International Journal of Energy Research. 2020;**44**(3). DOI: 10.1002/er.4883

[19] Pfeffer T. The evolution of information and communication technologies. In: Virtualization of Universities. 2012. pp. 5-20. DOI: 10.1007/978-1-4614-2065-1_2

[20] Wang T, Li G, Huang B, Miao Q. Spectrum Analysis and Regulations for 5G, 5G Mobile Communications. 2016. pp. 27-50. DOI: 10.1007/978-3-319-34208-5_2

[21] Directorate-General for Communications Networks, Content and Technology, Radio Spectrum Policy Group Secretariat, RSPG17-006 FINAL, 'Opinion on the Spectrum Aspects of the Internet-of-Things (IoT) Including M2M', Brussels, 2016

[22] Wicks A, Whiteford G. Value of Life Stories in Occupation-Based Research. 2003. DOI: 10.1046/j.1440-1630.2003.00357.x

[23] Faire J, Millan R, Ariel M, Bartalev S. Small satellites for space science: A COSPAR scientific roadmap. Advances in Space Research. 2019;**64**(8):1466-1517

[24] Duroc Y. From identification to sensing: RFID is one of the key technologies in the IoT field. Sensors. 2022;**22**(19):7523. DOI: 10.3390/s22197523

[25] Zeadally S, Siddiqui F, Baig Z. 25 years of bluetooth technology. Future Internet. 2019;**11**(9):194. DOI: 10.3390/fi11090194

[26] Huang K. On wireless local area networks [PhD thesis]. National University of Ireland Maynooth; 2010

[27] Trivedi D. Near field communication overview and applications. Stevens Institute of Technology; 2015

[28] Pons M, Valenzuela E, Rodiguez B. Utilization of 5G technologies in IoT applications: Current limitations by interference and network optimization difficulties—a review. Sensors. MPDI; 2018;**23**(8):3876. DOI: 10.3390/s23083876

[29] Worighi I, Maach A, Hafid A, Hegazy O, Van Mierlo J. Integrating Renewable Energy in Smart Grid System: Architecture, Virtualization and Analysis. Elsevier; 2019. DOI: 10.1016/j.segan.2019.100226

[30] Zheng J, Gao D, Lin L. Smart meters in smart grid: An overview. In: Green Technologies Conference. IEEE; 2013. p. 17. DOI: 10.1109/GreenTech.2013.17

[31] Benevides, da Costa D, Yang H. Grand challenges in wireless communications. Wireless Communications. Frontiers in Communications and Networks 1. 2020. DOI: 10.3389/frcmn.2020.00001

[32] Nižetića S, Šolić P, González-de-Artaza D, Patronod L. Internet of things (IoT): Opportunities, issues and challenges towards a smart and sustainable future. Journal of Cleaner Production. 20 Nov 2020;**274**:122877. DOI: 10.1016/j.jclepro.2020.122877. Elsevier

[33] Basso T, Hambrick J, DeBlasio D. Update and Review of IEEE P2030 Smart Grid Interoperability and IEEE 1547 Interconnection Standards. IEEE; 2012. DOI: 10.1109/ISGT.2012.6175748